*Stalking
the Blue-Eyed
Scallop*

Stalking
the Blue-Eyed
Scallop

by EUELL GIBBONS

with illustrations by
CATHERINE R. HAMMOND

with a foreword by
SHIRLEY KING,
author of Fish: The Basics

Alan C. Hood & Company, Inc.
Chambersburg, Pennsylvania

Library of Congress Cataloging-in-Publication Data

Gibbons, Euell.

Stalking the blue-eyed scallop.
Reprint. Originally published: New York:
D. McKay Co., c1964.
Includes index.
1. Shellfish. 2. Seafood. 3. Seashore
biology—United States. I. Title.
TX387.G5 1988 641.6'94 88-13399
ISBN 0-911469-05-2 (pbk.)

Published by Alan C. Hood & Company, Inc.
Chambersburg, Pennsylvania

10 9 8 7 6 5 4 3 2

Foreword

A TEENAGE Euell Gibbons strapped on a knapsack, took leave of his starving family and went into the mountains in search of food. It was the Great Depression, and his quest was born of stark necessity. He returned with an odd assortment of wild foods that kept the family sustained. His meager harvest of nuts, wild mushrooms and prickly fruit started within Gibbons an awakening that would become a lifelong fascination with nature. Over the years he foraged in fields, forests, streams and beaches, he dredged in tidepools, and he dipped, dunked and dragged for his dinner, all the while taking notes to improve his future harvests.

As he stood in a tidepool, a single splash brought Gibbons his first scallop. Opening it minutes out of the water, smelling its sea-sweet aroma, removing the plump, white muscle, passing it ever so briefly over an open fire and sitting down for that first taste of nirvana was surely the point of no return.

What was born of necessity became a lifelong adventure. Having had no success as a writer of fiction, Gibbons was finally convinced by his editor to record his extensive knowledge of wild foods. Beginning with *Stalking the Wild Asparagus*, he devoted the rest of his life to writing his celebrated foraging handbooks with their delicious, simple recipes.

Yes, scallops do have blue eyes. *Stalking the Blue-Eyed Scallop* is a cookbook, nature guide and remembrance that pulls you in like a riptide. It is particularly illuminating about all the fine seafood we can gather on our shores, with beautiful descriptions of their habitats and behavior, and not least, recipes for some of the most unique and savory dishes you will ever enjoy.

SHIRLEY KING, 1997

Shirley King is a professional chef and author of numerous cookbooks and articles including the classic, *Fish: The Basics* (Simon & Schuster, 1990; Rev. Ed., Chapters, 1996). She appears regularly on radio and television cooking programs including "Fish Talk with Shirley King," a monthly visit with Arthur Schwartz on WOR Radio.

Acknowledgments

WRITING a book of this kind is a humbling experience. It is only when one starts putting such lore on paper that one realizes how much is owed to others and how little has come from independent research. My love affair with the shore goes back more than thirty years and during that time I have consulted countless books on this subject. Some of the outstanding ones, such as Rachel Carson's *Edge of the Sea*, Ed Ricketts' *Between Pacific Tides*, and R. Tucker Abbott's *American Seashells*, leap instantly to mind, but there are dozens of others from which I have taken what I needed at the time and then, ungratefully, forgotten. To all these authors, remembered and unremembered, I am belatedly but sincerely grateful. Nor must I forget the many patient and helpful librarians who assisted me in locating these books. Deserving special mention are Miss Margaret Greenwald, Librarian at the Academy of Natural Sciences in Philadelphia, and her staff, for the uncomplaining efforts they devoted to making this book a possibility.

Many other staff members at the Academy of Natural Sciences were helpful and I'll never forget the friendly patience with which they extended aid to this bothersome, upstart author. Thanks go especially to Dr. R. Tucker Abbott, Chairman of Malacology, and author of several fine books on seashore life, for reading and correcting some of the chapters dealing with mollusks. Also to Dr. Alfred E. Schuyler, Chairman of the Botany Department, for free use of the Academy's excellent herbarium and for correcting some of my botanical errors. There have been many additions and changes in the manuscript since these two men saw it, and any mistakes you find are mine, not theirs.

I must also thank two authorities at Bucknell University—Dr. Wayne E. Manning, Professor of Botany, and Dr. Roy C. Tasker, Professor of Zoology.

Not nearly all the lore I have set down here came from books. It was an old Makah Indian in the State of Washington who showed me the best way to shell and eat a Dungeness Crab; a little naked Hawaiian boy taught me how to lasso a Mantis Shrimp; a very intelligent Negro lad in Georgia demonstrated how to extract a Stone Crab from its burrow without getting pinched; an old lobsterman in Maine showed me how to catch fish with Dog Whelk bait. To these and to the many other chance seaside acquaintances, some perhaps not so colorful but all helpful, I extend my thanks.

There are so many others to whom I am indebted: Catherine Hammond, artist, who not only produced the fine line drawings which illustrate this work, but also shared with me her favorite beachcombing spots along the Massachusetts coast, and helped to capture some of the creatures described in this book; Annette Carter, naturalist, and nature writer for the *Evening* and *Sunday Bulletin* of Philadelphia, who helped on several seaside expeditions, waded tidal streams, slogged through mud flats, and bravely sampled many experimental seafood dishes. She kindly consented to read portions of the manuscript, and her helpful advice and friendly criticism have been greatly appreciated. On one of our salt-marsh expeditions she took the color photo which appears on the jacket of this book. Then there is Mrs. Sachiko Presser of the library staff at Susquehanna University who not only aided the library research, but contributed lore on edible seaweeds and Japanese seafood dishes that came from her own experience. There is Eleanor Rawson, of David McKay Company, who is all an editor should be and who understands both how to curb my wild excesses and how to spur me out of periodical slovenliness. Especially there is my wife, Freda, who spurs me on with love and encouragement and protects me from interruptions while I am writing. She has kept us orderly, well-fed, and socially acceptable despite my unusual hobby of experimenting with strange, exotic, and often barbaric foods.

But why go on? A book of this kind is always a group effort, no matter whose name appears on the title page. To the large number of persons, named and unnamed, who have contributed to the writing of this book, I am profoundly grateful.

EUELL GIBBONS

BEAVERTOWN, PA.
MARCH 17, 1964

To FREDA,

my favorite beachcombing partner,
a good companion, patient critic,
tender comrade, and incidentally
MY WIFE

Contents

Illustrations

xiii

*Stalking
the Blue-Eyed
Scallop*

1. How to Cook a Sea Serpent

WE were three couples who had rented a beach cottage on Delaware Bay for the first two weeks in September. Summer lingered that year, and while our wives enjoyed swimming in the bay and sunbathing on the beach, we three men concentrated on fishing. Out in the bay and in the many nearby tidal creeks we caught dozens of fish of several species and any number of large Eels.* At my insistence we also went Oyster tonging, gathered Mussels and Periwinkles from a rock jetty, and caught bushels of large Blue Crabs in ring nets and folding traps.

My inordinate and abiding interest in food was well known to my companions, so I was unanimously elected dinner chef for the duration of the holiday. I know that cooking can be an onerous task to those who dislike it but still have to do it regularly, but to me it is an exciting hobby, so I didn't feel at all imposed upon. My interest in food is not that of the conventional gourmet. I have never had the means or the inclination to concern myself only with fine restaurants, expensive hotels, or skilled chefs. I am as fascinated by the origin, preparation, cooking, and serving of food as I am by its final

* Capitalized names refer to subjects that receive more or less extended treatment in this book.

1

consumption. Especially am I thrilled when I can garner good food-stuffs from fields, forests and streams, seabeaches, bays and tidal creeks, then turn these rude ingredients into civilized dishes that can be appreciated by the most fastidious epicure. Then I can feel that I'm being original and creative, and not merely completing a job that has been more than half done for me by the processors of commercial foodstuffs.

We not only used the glorious seafoods so abundantly available, but also scoured the nearby woods, fields, and beaches for herbs, wild vegetables, and flavoring materials to supplement the offering of the sea. Having all these unusual ingredients to work with, as well as the commercial food we could buy at the beach store, I tried to make every dinner a feast. We had minutes-fresh saltwater perch filleted with the skin on and fried to a crispy gold; Striped Bass larded with bacon and stuffed with Sour Sorrel and rice, then baked in the cottage oven; weakies cooked like mountain trout and tasting even better. The freshly tonged Oysters became *Huitres aux champignons frais*, Oysters Rockefeller, and exotic stews flavored with leaves of Bayberry. The crabs we caught by the dozen became Deviled Crab, baked in their own shells, Crab Cakes, Crab Louis. One night, our dinner was just a mountain of crab, boiled in seawater with crab-boil spice and Bayberry leaves, cleaned, cracked, and served in the shells. We tackled the hot crab with nutpicks, dousing the meat with Melted-Butter Sauce and eating it with tart Beach-Plum Jam. With this we had pink champagne, taken in small sips between bites to freshen the taste buds for the next delicious mouthful. Such a meal required a plastic tablecloth and bibs on all diners, but it was a memorable feast.

My companions had never eaten Eel, Mussels, or Periwinkles, but when I served sections of fat Eel stuffed with a smart-tasting Tabasco dressing and baked in buttered foil, they quickly lost their prejudices. A dish of French Mussels seasoned with thyme, shallots, and Chablis met unanimous approval. A delicate Periwinkle Omelet (my version of *Omelette à l'Arlésienne*), studded with chopped chives, was greeted with cheers.

One of my companions was a Southerner who had been endowed with an appreciative sense of taste that made cooking for him a sheer joy. On a day when our stay was about over, we three men were out in the boat, having a last go at the fish, and he remarked that the only thing needed to make our vacation perfect would be for one of us to catch a sea serpent. When asked why he wanted a sea serpent he said, "I'd just like to see how Euell would cook it, and by golly, I'd like to taste the dish he would make, too."

I have never had an opportunity to try my hand at sea-serpent cookery, but I do deplore the lack of knowledge and the unreasoning prejudice that cause us to ignore such epicurean foods as Mussels, Periwinkles, Eels, Whelks, Angel Wings, Pen Shells, and Surf Clams. There is great delight in seeking out old recipes for unusual seafoods and in devising new ones. A great field for fascinating experiment lies among the edible seaweeds with their valuable offering of essential minerals, and among the tangy, vitamin-loaded plants that grow near the sea. The collection, preparation, and enjoyment of nature's seaside bounty is the activity that gives most meaning to all my expeditions along the shore.

Do I, then, look on the shore only as a source of good food? Not at all. I love the broad expanses of sandy beach, the pounding of the surf on a rocky shore, the saltwater bathing, and the rippled dunes as well as the next person. I am fascinated by the waterworn shapes of driftwood, the iridescent shells, and all the strange flotsam cast up by the ocean waves. My interest in the myriads of creatures that inhabit the life-crowded strip between the tides and my appreciation of their beauty are in no way diminished because I know which of them can minister to my needs—rather the opposite.

Appreciative acceptance of the bounty so graciously offered relates us to seaside nature in a new and deeper way. The most sacred rites of higher religions involve ceremonial eating and drinking. Surely the way to approach true communion with the sea is the grateful reception of this free gift of food that has never been gathered for gain or sold at a profit, preparing it with the loving care that lifts cooking from an irksome task to a fine art, then eating it with a reverent awareness, not only of its taste, texture, and aroma, but also its very nature and origin. Such sacramental food nourishes the soul as well as the body.

All life originally came from the sea, but few people realize the extent to which the sea nourished the childhood of our race. We are accustomed to think of our primitive ancestors as inland cave dwellers and mighty hunters before the Lord, and some of them were, as archeology proves. But there is ample evidence that a great many of them were seaside dwellers, fishermen, and eaters of shellfish. Some of the most astonishing relics of ancient man are the huge shell mounds, or "kitchen middens," found on seacoasts in many parts of the world. The extent and depth of some of these mounds are truly amazing. They cover many acres and rise up like young mountains, each mound containing thousands of cubic yards and literally billions of individual shells. When one recalls that each of these shells was opened by a pair of human hands and that the succulent food morsel that lived in each of them once disappeared between a pair of human lips, one stands in awe at the thought of all the good eating that occurred on that spot.

When modern archeologists excavate these shell mounds they always find the remains of ancient campfires, soot-blackened hearthstones, and the stone and bone tools lost or discarded by Pleistocene man. Occasionally they find the cooked bones of land mammals and upland game birds, showing that our forebears sometimes enjoyed a haunch of venison or a roast grouse, but their chief dependence was on the products of the sea, for 95 percent of the mass of these prehistoric mounds is made up of the discarded parts of edible shellfish.

Even after man achieved civilization, he did not forget his marine heritage. Those magnificent gourmets, the ancient Greeks and Romans, featured many kinds of fish and shellfish at their sumptuous banquets. When man goes foraging in that mysterious zone between the tides, he is not raiding an alien or hostile territory; he is returning to his ancient racial home and a noble one.

As we walk along that wet strip of land that belongs neither to the shore nor to the sea, we feel a strange kinship with the creatures that are today engaged in the ages-slow process of coming ashore. The scuttling dune crab almost has it made. Already he prefers to live above the highest tides, and, when frightened, runs toward his burrow on the shore rather than toward the sea. However, he can't wander too far inland, for in his reproductive cycle he is still bound to the sea, and most of his food comes from the wet beach at ebb tide.

The Rough Periwinkle would seem to be further along the road toward becoming a land creature. He has learned to bring forth his young alive and in his own image, but he is still bound to the upper shore by a preference for seafood and a strange vitality cycle which, like the tides, is linked to lunar phases.

Living among the upper rocks that are reached by the sea only during the spring tides, the Rough Periwinkle exhibits an energetic burst of activity during these fortnightly visits of the waves, then is dull and inactive during the neap tides. It would be a mistake to assume that the movement of the tide *causes* these changes. A scientist once placed some of these interesting creatures in an aquarium and observed them for many months. In an environment kept constantly the same, these amphibious little snails kept exhibiting the same alternations of activity and quiescence, and their periods were exactly timed with the fortnightly springing and quieting of the tides.

One reasonable hypothesis is that these cycles are caused, not by the tides, but by the same forces that cause the tides. When the moon is new, and again when it is full, the solar and lunar orbs are pulling in line with one another, and not at an angle, and these are the times of maximum tidal ebb and flow, the spring tides. It has nothing to do with the time of the year. Certainly the combined gravitational attraction of our two chief heavenly bodies also raises an infinitesimal tide on the microscopic sea contained in each living cell of the Rough Periwinkle. This tiny, internal tide must be the trigger that awakens this fascinating little gastropod to his semi-monthly periods of intense activity.

No creature has absolutely completed the process of coming forth from the sea. Even man betrays his marine origin in many ways. Does not the human foetus grow gill slits at one stage of its development; is not an analysis of the liquid parts of human blood almost identical with that of seawater; and do not our females ovulate on a lunar cycle, in time with the tides? When the moon is new and thin, and again when it is round and full, the lined-up planetary forces must also affect the liquid parts of our own cells, and perhaps we, too, feel a pull toward the sea when the spring tides begin to ebb.

But aren't the shores all taken up by resorts and boardwalks, or else ruined by industrial pollution? Nonsense! There are 54,000 miles of tidal shoreline in the United States south of Canada. Alaska has 26,000 more miles. Add to these the marvelously indented and island-studded coast of British Columbia, the winding shores of the Maritime Provinces of Canada, and the nearby shores of Mexico, and you have at least 150,000 miles of saltwater shoreline that is accessible to Americans. Of this, only a few miles are occupied by resorts and boardwalks. Far too much of it has been ruined by industrial pollution, but subtract all that man has rendered unfit for the kind of recreation advocated here, and there is still a length of tidal shoreline that would reach more than five times around the world. Let's stop mourning for the good old days. We are largely living in them still, so let's get out and enjoy those thousands of miles of shoreline that are left to us.

The methods, techniques, and recipes given in this book can be applied by any seaside dweller, summer cottager or seashore camper. If you live inland, one of the finest ways to experiment with this method of enjoying the shore is the family camping trip. A camp trailer, one of the bus campers now offered by many of the automobile companies, or just a tent and some camping equipment, will open this wonderful world to you and your family. Once the initial cost of equipment is past, such vacations are not expensive. The rental of camp sites in state parks or private camp grounds is nominal, and after you have mastered the techniques of gleaning a large part of your food from the sea, the shore, and the surrounding woods, you will find a family can take such a camping vacation almost as cheaply as they could have stayed at home.

Attempting to take a camping trip along the seashore without such a program of merging with nature and learning to accept her offerings is likely to prove disappointing. The older children become bored and drift into nearby resorts to eat hot dogs, order carbonated drinks, go to the movies, or play pinball machines. When this happens, the children are not to blame. Getting back to nature, when we know nothing about nature and are not interested enough to make a closer acquaintance, is like going through an art museum where all the pictures are turned to the wall. Where there is "Little we see in Nature that is ours," nature soon becomes damnably boring.

How different things are when a family sets out really to live in contact with nature, gleaning as many of their needs from natural sources as possible. I have never known a child who could not be interested in an attempt to live off the land for a time. Food is elemental, and when a child is given such immediate and compelling reasons for learning nature's secrets, he throws himself into the study with surprising enthusiasm. Such a holiday can be a very valuable educational experience, with children learning more than they ever would in a comparable time in school where motivation is poor and the rewards are, to a child's sense of time, impossibly distant.

Gathering even a small part of one's food from the wild not only brings new and useful knowledge, but changes attitudes. A child who has participated in food-gathering in such a basic and original way will never be indifferent to food again. I have seen several serious feeding problems solved by introducing the child to this kind of foraging. To gain the maximum benefit from this activity, it must be approached with the right spirit. We must never imagine ourselves engaged in the conquest of nature, nor think that we are making the sea and shore stand and pay tribute. Nature is no enemy to be conquered and subjected to our personal use, but a friend with whom we can and must cooperate. When we dig a succulent clam or gather a tender wild vegetable we are not being clever raiders, but are merely claiming an ancient inheritance. If nature is approached with humility and respectful love, she will teach us and feed us, for we are not above or alien to nature, but part of her—in fact, her favorite children.

The parents who would take their families on such interesting adventures need not be graduate biologists, knowing every seashore creature and woodland plant by its Latin name. Take this book with you and select a few texts on botany, seashore life, and fishes, and do your studying with the specimens right before you. How quickly we can learn when our dinner depends on the knowledge, or when it means, at least, the difference between a meal of fresh wild food and a tasteless mess out of tin cans.

To keep your diet well balanced and your experience well rounded, do not neglect the wild-plant food mentioned in this book. There are fewer edible species of plant growing along the seashore than you will find farther inland, but these few species are often abundant as individuals and very fine in quality. This apparent poverty of species along the shore need never worry the seaside camper, for one has only to walk over those dunes to reach the great wealth of edible wild plants that grow inland. Many of these so-called inland plants are, themselves, often found growing near the shore. The common cattail often penetrates brackish bays, and its roots, young stalks, bloom spikes, and pollen all furnish food dear to the knowledgeable forager. Purslane, good as a cooked vegetable or raw salad, is often found thriving on the landward side of the dunes. Cranberries, blueberries, huckleberries, blackberries, and wild grapes all flourish within sound of the sea. Around Chesapeake Bay I have seen Juneberry bushes and chokecherry trees growing just above the tide line, so loaded they could hardly hold up their burden of ripening fruit.

Directions for recognizing, gathering, and making use of these inland plants will not be given in this volume, for they are covered in great detail in my earlier book, *Stalking the Wild Asparagus*. That book and this are intended to be companion volumes, and the seaside camper who wishes to make maximum use of the wild foods he can find growing near almost any camp site will include both books in his camping equipment.

The foraging vacation is not my own invention; it has been practiced in America for hundreds of years. The Lenni-Lenape, a pleasure-loving Indian tribe of the Delaware Valley, made regular expeditions to the Jersey shore for just such summer fun. This is not archeological conjecture, but a matter of history, for these annual excursions continued long after the coming of the white men. As soon as the fields had been prepared and the corn, beans, and pumpkins planted, the Indians formed congenial groups and traveled across New Jersey to the shore. Temporary wigwams were soon built amid much shouting and laughter as the arriving groups renewed old friendships and made new ones.

What a happy hunting ground was this land of plenty! The men caught Lobsters and many kinds of fish with net, trap, and spear. Some waded the bays treading out Quahogs while others went to the outer beaches for Surf Clams, Mussels, and other ocean shellfish. Young people went crabbing or picked berries in the nearby woods. Any amount of Oysters could be had for the gathering. Children hunted eggs in the seabird rookeries or set snares for quail, grouse, wild turkey, and rabbits. They ventured into the salt marshes and captured dozens of the delicious diamond-back terrapins, like those that now bring fantastic prices in gourmet restaurants.

In the evenings there were campfires, Clambakes, Crab Boils, and general feasting, flirting, and fun. There were Indian dances, pageants, and powwows. Here occurred the annual parade of virgins, a sort of coming-out party for maidens who had reached the age of eligibility, an ancient precursor of the Miss America Pageant now seen on this same shore. With food so abundant, the Indians were like happy animals, consciously overeating in a deliberate attempt to put on an extra layer of fat against the hard times of the coming winter.

An early poet, William Wood, published a quaint verse in 1639, giving us a picture of Indian life that reveals the major role played by seafood in the cuisine of these carefree people.

> "The dainty Indian maise
> Was eat with clamp-shells out of wooden trays,
> The luscious lobster with the craw-fish raw,
> The brinnish oyster, mussel, periwigge,
> And tortoise sought by the Indian squaw,
> Which to the flats dance many a winter's jigge,
> To dive for cockles and to dig for clams,
> Whereby her lazy husband's guts she cramms."

These gay times and this good eating can still be enjoyed if one picks the right stretch of shore. I doubt that we will persuade our modern wives to work like Indian squaws to procure these dainties, but there are localities where all the delicious seafood a family can eat can be secured by an hour or so of foraging at low tide each day. This leaves many hours for other seaside pleasures such as swimming, boating, shell collecting, surf fishing, building sand castles, or just plain loafing in the sun.

2. Gather Your Own Oysters

(*Crassostrea virginica*)

According to experts, the oyster
In its shell—or crustacean cloister—
    May frequently be
    Either he or a she
Or both, if it should be its choice ter.

Berton Braley

A LTHOUGH the naturalist in me cringes at the use of the word "crustacean" in connection with a mollusk, the above limerick points up a memorable fact about the perfectly fascinating love life of the Oyster. Most Oysters start life as males; then, as middle age approaches, they decide to become females, and they do! Therefore, the Oysters we eat are nearly all the offspring of old mammas and young daddies.

A fully mature female Oyster (and fully mature Oysters *are* females, of course) can produce up to sixty million eggs in a season. Naturally all these ova do not develop into succulent bivalves of the size we relish on the half shell. If they did, the whole world would soon be covered with Oysters, and that would be too much of even as good a dish as they are.

The aged females simply entrust their millions of unfertilized eggs to the tidal currents that flow about them, while the young males do the same with their even more copious sperm. In the vastness of the tides these microscopic beings find one another, and the eggs are fertilized. These promptly hatch into great multitudes of tiny, top-shaped larvae, the free-swimming *trochophores*. In a few days these metamorphose into the so-called *veliger* larvae, and at this stage they start building shells about themselves and looking for a place to settle down.

Oysters do not thrive in the concentrated brininess of the ocean, so for likely homes they must seek the bays, estuaries, and tidal creeks where freshwater streams have diluted the seawater to the requisite brackishness. The unattached larvae instinctively know this, and when the tide is racing out they settle to the bottom and try to hold on; then, when the tide turns and sweeps inland, they let go and float with it. A very few of the many millions eventually come to rest on suitable rough surfaces such as stones or dead shells. These lucky ones anchor themselves immovably for the rest of their lives and concentrate on growing into Oysters of edible size.

Obviously you can't count your Oysters before they hatch, nor after, either, for that matter. It seems the destiny of the Oyster to furnish delicious food for a broad spectrum of life, from tiny copepods that consume billions of the eggs and larval young to massive gourmets who enjoy Oysters on the half shell or *Huitres aux champignons frais*. Crabs and bottom fish crush the developing shells and eat the young Oysters inside. Mature Oysters even consume their own eggs and larvae unconsciously, if these happen to get sucked into their feeding siphons. One would think that the ones who survived to construct rock-hard shells about themselves would finally be safe, but not so. Starfish, by sheer strength and patience, force these shells open, then literally turn inside out, poking their everted stomachs between the shells and digesting the Oyster *in situ*. The oyster drill with its filelike tongue actually licks a hole through these hard shells large enough to insert its mouth, which is the biological equivalent of a soda-fountain straw, and then it simply sucks out the semiliquid creature.

Finally, if all these hazards are safely passed, the oysterman comes along with his dredge or tongs, and the Oyster ends up on the plate of some appreciative epicure. The Oyster seems to have no defense against this multitude of predators except its shell, and that doesn't always defend it. As a species, the Oyster only survives because of its amazing fecundity.

It is this marvelous fruitfulness of the Oyster that guarantees that there will always be young Oysters settling down and growing to maturity in places away from the cultivated beds of commercial oyster-men. There are a great many more of these "wild" or uncultivated Oysters around than most amateurs think. Some states even set aside certain tidal river systems or hunting-and-fishing grounds where the Oyster rights are reserved for the general populace. I have gathered wild Oysters in New York, New Jersey, Delaware, Maryland, and Virginia, and I'm sure uncultivated Oysters exist in many more coastal states. If we neoprimitive wild-food gatherers made our appreciation of these unclaimed Oysters known in the right places, there would soon be more areas available to us, for, as in the matter of hunting and fishing, the states have long since learned that it is far more profitable to attract tourists and sportsmen than it is to lease fishing or hunting rights to private exploiters.

It is easy to determine whether or not there are Oysters on the bottom of the tidal stream, bay, or gunkhole you frequent. If there are Oysters about, almost every dead shell or rock below low-water mark will show little nacreous spots, smooth and light-colored. This is called "spat," and each of those little pearly spots is a developing young Oyster with the translucent beginnings of a shell.

I have gathered half a canoe-load of Oysters without getting my feet wet, by just drifting along one of the tributaries of the Maurice River, in New Jersey, at low tide and pulling the Oysters from their anchorage at low-water line. In Delaware, I rubber-booted up and down the small streams in a salt marsh and collected half a bushel during one low tide. In Maryland, I fished a tidal stream until the low tide ruined my fishing, then stepped barefoot into the stream and gathered enough Oysters for our camp dinner.

These easy ways of gathering Oysters are the result of rare luck or unusual conditions; to get Oysters regularly requires more work. I know tidal creeks where I can wade in at low tide, locate the Oysters with my feet, then duck under and bring them up. Unfortunately, Oysters are only in season during months with an "r" in them, and these are pretty cold months. It is fun to gather Oysters in this manner during the warm days of September, but after the water gets too cold for swimming, you're going to need a pair of oyster tongs.

These are a pair of gigantic pliers with wooden handles and stout iron baskets for jaws. They can be purchased near commercial oystering centers, or a local blacksmith can make you a pair. They come in various lengths, but I have found an 8-foot pair about right for use when the tide is low. They are used from a boat and one simply opens them up, shoves the jaws against the bottom, works them together, then raises the Oysters trapped inside. Sounds easy, but just wait until you see how many times those baskets come up with nothing but dead shells in them.

Those who dread the job of opening Oysters do not deserve the epicurean morsels within. It is neither a difficult nor a tedious task once the technique is learned. Look at an Oyster edgewise, and you will see that one shell is rounded and the other somewhat flat. The end of the Oyster that was attached to its underwater mooring is the hinge end; on the other end the shells grow on beyond the animal rather wildly, and these protruding ends are called the lips. Place the Oyster on a firm surface, flat side up, and break off enough of the excess lips so you can see an opening large enough to insert a knife blade. I do the breaking with a heavy kitchen knife, a hammer, or a pair of nippers. Slip a thin knife into the opening and slide it back against the flat side of the shell until you sever the adductor muscle. This relaxes the Oyster's tight hold, and you can pull the flat side of the shell off. Now slide your knife under the Oyster and cut its moorings from the rounded side of the shell, and it is ready to serve on the half shell or to be removed altogether and cooked in any number of delicious ways.

Oysters served in the rounded half of the shell, whether raw or cooked, are fancy fare and deserve decoration. To make raw Oysters on the half shell attractive, bed the shells in a plate of cracked ice and garnish with sprigs of watercress and bright wedges of lemon and tomato. Pass cocktail sauce to those who wish more seasoning.

Oysters Barbecued with Bacon are a taste thrill for the seafood lover. Put half an inch of rock salt in the bottom of a bake pan and bed in it the rounded halves of the shells containing the Oysters. The salt serves to keep the shells in position so they won't tip over and spill out the liquor. Sprinkle the top of each Oyster with fine bread crumbs, add a dash of paprika, and then cover each Oyster with thinly sliced bacon. Put in the broiler about 3 inches below the flame and broil until the bacon is crisp, about 15 minutes. Serve them sizzling hot and, instead of the usual lemon wedges to squeeze over them, try serving these Oysters with thin slices of tangerine. It makes something very special of this dish.

While camping in autumn or spring we eat most of the wild Oysters we find, raw, fried, or stewed. Fried Oysters can be done to perfection over a camp stove or campfire if enough care is taken. Shuck the Oysters, drain them, and pat them dry between paper towels. Beat together 1 egg, 1 teaspoon monosodium glutamate, and a dash of freshly ground black pepper. Yes, we take our pepper mill on camping trips, but you can put a few peppercorns in a scrap of cloth and pulverize them between two stones if you want to. Put some saltines in a plastic bag and crush them to very fine crumbs. You will need about 2 crackers per large Oyster. Dip the Oysters in the egg mixture, then into the crumbs, and they're ready to fry. If the Oysters are small, hold two together and dip them into the egg, then into the crumbs, then back into the egg, and back into the crumbs again. This holds them together and they fry without becoming too hard in the center, as small Oysters tend to do when fried singly. At home I usually fry Oysters in a wire basket submerged in hot fat, cooking them about 3 minutes, but in camp I use a little bland cooking oil in a frying pan and turn the Oysters after about 2 minutes, cooking them a little less on the other side.

When making Oyster Stew in camp, I cook a pint of Oysters in their own liquor until the edges begin to curl. Then I add 1 quart of milk, a little black pepper, a teaspoonful of monosodium glutamate, and 2 tablespoons of butter. This is heated just to the boiling point and served with sesame-seed crackers.

When I wish to serve a really swanky Oyster Stew, I sauté a pint of Oysters with their liquor in 2 tablespoons of butter and add a leaf from a Bayberry bush. When the edges curl I add 1 pint of milk, 1 pint of cream, ¼ cup cooking sherry (not beverage sherry, but *cooking* sherry), ¼ teaspoon nutmeg, 1 teaspoon monosodium glutamate, and a dash of black pepper. When this is all heated through but not boiling, I remove the Bayberry leaf and serve the stew hot with a French loaf that has been sliced, buttered, and heated in the oven.

If you'd like to put on even more dog with the Oysters you have gathered with your own hands, I'll give you my recipe for *Huitres aux champignons frais*. Wild mushrooms are best, pasture mushrooms, puffballs, or lepiotas, but if these are unavailable or you aren't sure of recognizing them, fresh mushrooms from the store will do; you need half a pound. Simmer 1 pint of Oysters in their own liquor until the edges curl, then drain. Melt 2 tablespoons butter in the top of a double boiler, then stir in 1 tablespoon flour. Let cook for 10 minutes to remove the raw, starchy taste from the flour, then add the Oyster liquor, 1 cup milk, and ½ cup rich cream. Cook until thick. Smother the mushrooms in a little butter until they just start turning yellow, then add the mushrooms and the Oysters to the cream sauce. Add salt, pepper, and monosodium glutamate to taste. Pour over toast rounds and sprinkle with finely minced chives and parsley. Serve with a good white wine.

COON OYSTERS
Ostrea frons

From the Carolinas southward, one comes on seaside groves of that widespread and interesting tree of muddy, saltwater shores, the mangrove. It actually grows between the tides with its feet in saltwater and is easily recognized by its tangle of arched prop-roots, which are above water at low tide. On these aboveground prop-roots one finds the Coon Oyster in great clusters and bountiful abundance. The Coon Oysters get their name because raccoons go out at low tide and feast on them, and when I am in the South I give the 'coons some competition. Coon Oysters are small and tedious to shuck, and it is a muddy job gathering them, but don't let anyone tell you they are not good.

Only about 2 inches long, and as wide as it is long, this little Oyster is easily recognized by the radial pleats or folds that characterize its shell, giving it correspondingly sharp folds at the valve margins. In color it is usually purplish red outside and translucent white inside. The Oysters are fastened to the roots by clasping projections of their shells, but are not difficult to pull loose. Be sure to gather plenty, for it takes a large quantity of Oysters in the shell to yield a serving of shucked Oysters. The shucking job is much the same operation as with the large Oysters, except that the yield is smaller. Break away enough of the lip so a thin knife can be inserted, sever the muscle that is located well back toward the hinge, and the Oyster is at your mercy.

Coon Oysters can be used in the same ways as the larger kind, except that when you fry them you should hold several together and dip first in beaten egg, then in cracker crumbs, then back in the egg and into the crumbs again, before you consign them to the hot fat. These fried clusters of Oysters are very good. In stews, soups, and other Oyster dishes, Coon Oysters are excellent. Try some Scalloped Oysters in individual ramekins with these little fellows. For each serving you will need ½ cup shucked Coon Oysters with their liquor. Butter as many individual baking dishes as there are people to serve. In each one, place 2 tablespoons cracker crumbs, then ¼ cup Oysters, then another 2 tablespoons of cracker crumbs, and another ¼ cup of Oysters. Mix together 4 tablespoons cream, ⅛ teaspoon salt, a dash of pepper, a dash of nutmeg, a few drops of Worcestershire sauce, and one drop of Tabasco. Pour this over the mixture, top with bread crumbs, and dot generously with butter. Bake in a 400° oven for 20 minutes and serve.

OLYMPIA OYSTERS
Ostrea lurida

Beachcombers along the West Coast have at their disposal the finest-flavored Oyster of them all, the small Native Oyster or Olympia Oyster. It reaches a maximum size of about 3 inches long and 3 inches wide; the shell, generally fan-shaped, is, like many other Oyster shells, very variable and often misshapen. The Olympias cluster on rocks in sheltered tidal water and are abundant in all suitable habitats from Sitka, Alaska, to Cape San Lucas at the tip of Lower California. They are little used over much of their range, but this neglect is entirely due to the tediousness and time involved in gathering and shucking them, and not to any dislike of their flavor, which is universally recognized to be very superior. It takes approximately two hundred of these little Oysters to make one pint of shucked Oysters, solid pack.

Despite this small yield, the Olympia Oyster is so excellent that appreciative gourmets are willing to pay a high enough premium on this delectable shellfish to make it profitable for tidewater farmers to cultivate it commercially. These oyster-farmers know they have a superior product and make no attempt to compete with large, ordinary Oysters in price. They charge enough to pay for all that extra labor that is involved, and still have no difficulty in disposing of all they

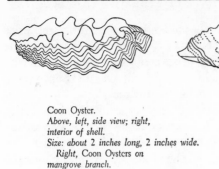

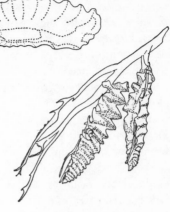

Coon Oyster.
Above, left, side view; right,
interior of shell.
Size: about 2 inches long, 2 inches wide.
  Right, Coon Oysters on
mangrove branch.

Olympia Oyster. *Below, left, exterior view. Right, interior of shell.*
Size: about 3 inches long, 3 inches wide.

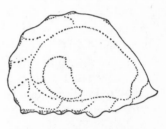

Quahog. *Left, exterior view. Right, interior of shell.*
Length: about 4 to 5 inches.

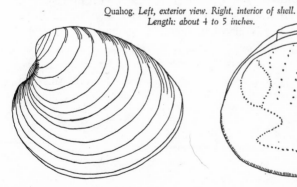

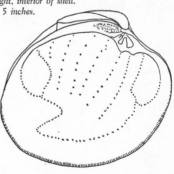

can produce. I have very willingly paid up to three times the price of large Oysters for these tiny delicacies and never felt cheated in the least. However, this was before I discovered that I could gather all of these fine little Oysters I could use from the countless bays along the much-indented coast of the Pacific Northwest.

Many years ago, four of us went for a week's hiking-camping trip along the shores of a thinly settled part of Puget Sound. We carried blankets, shelter halves, cooking utensils, and food for the first day in our packs, and planned to buy other food from stores along the way. However, from the first day we found a great abundance of Basket Cockles, Horse Clams, Blue Mussels, Butter Clams, Dungeness Crabs, Rock Scallops, and best of all, a great plenty of Olympia Oysters. All this wonderful seafood almost directly from the cold salt-water to the pot or plate was so delicious that we completely neglected other foods and arrived at our destination a week later without having even finished the supply of food we carried to see us through the first day. At wayside stores we had bought only butter and condiments to use in preparing our seafood meals. We found Hot Clam Nectar so delicious and bracing that we didn't even make coffee. Although we ate Olympia Oysters every day, we varied the ways in which we prepared them and didn't tire of them at all.

Olympias are the finest possible Oysters for eating raw, and they are delicious prepared in any of the ways you would fix any other Oysters. For frying they must be dipped and crumbed in clusters as directed for Coon Oysters, or they will cook too hard in the center by the time the outside is browned. Better than frying Olympias is to sauté them just as they are taken from the shells, with no coating of crumbs. Melt 2 tablespoons butter over a low fire. Slowly cook 2 cups shucked Olympia Oysters just until the edges begin to curl. Add ½ cup cream, 1 teaspoon chopped parsley, ½ teaspoon salt, and a dash each of black pepper and Tabasco sauce. Serve on toast rounds or on the corn-cream pancakes described on page 100. To vary this dish, substitute crushed anise seed, dill seed, or celery seed for the chopped parsley.

3. Quahog or Littleneck:
THE WAMPUM CLAM

(*Venus mercenaria*)

ALSO called the Round Clam, Hard Clam, and when very young, the Cherrystone, this is the best-known edible clam from southern Canada to Florida. The Indians appreciated these clams long before the white men arrived, and they taught the Pilgrim Fathers how to gather and prepare them. The word "Quahog" is the white man's attempt to pronounce the Indian name for this valuable bivalve.

The Quahog is a roundish clam, and the full-grown ones may be 4 to 5 inches long and a little less across. In the South they grow even larger. The color is grayish white outside and creamy white inside, usually with one edge deep purple and sometimes with other purple splotches. These purple parts of the Quahog shells furnished the Indians with the material from which they made the most valuable wampum, worth twice as much as the ordinary white kind made from whelk shells. This bead money had a definite monetary value in the European coinages used in early America and was sometimes used as a medium of exchange between whites as well as among its Indian coiners. When an early settler tendered a golden guinea in payment for some low-priced item, he was likely to receive white wampum, purple wampum, English shillings, Dutch guilders, and Spanish pieces-of-eight in his change.

The Quahog is seldom found on the sandy, surf-pounded beaches with the Hen Clam, but prefers the somewhat protected mud-and-sand shores of estuaries and inlets. Here, where enough mud has mixed with the sea sand to give it a firm consistency and a blue-gray color, the Quahog is often found in company with the Razor Clam, and I have seldom gone Razor Clam digging that I didn't also secure some Quahogs. They are also found on the more muddy bottoms of tidal streams, if there is little admixture of fresh water, and here one finds them in company with Mussels and Oysters.

When hunting Quahogs on a mud-and-sand beach at low tide, the proper tool is a clam rake. The Littleneck, as its name implies, has very short siphons, which force it to lie just below the surface where it is easily located and raked out. Some foragers disdain even a rake when hunting this clam and search it out with their toes. When one clam is found, search the sand near it, for, while one does find solitary Quahogs, they tend to be grouped together in beds.

The old Indian method of "treading out clams" works well in shallow tidal runs, if you are not afraid of a little mud. Just wade along at low tide feeling through the mud with your feet until you encounter a clam, then pick it up. The human foot is quickly trained to recognize the rounded smoothness of the Quahog. I have sometimes gone clamming in water that was waist-deep, using a short pair of oyster tongs to lift the clams my feet located. When engaged in this underwater clamming, I always tow a basket or box behind me with an inflated inner tube around it to keep it from sinking when it gets loaded with clams. If you are willing to brave mud and water, you will usually be able to find plenty of clams, for then you get out where they have not all been taken.

How you eat a Quahog depends largely on its size. The little Cherrystones, about 1¼ to 2 inches in diameter, are delicious just opened and eaten raw with cocktail sauce or melted butter. When very hungry I have eaten as many as three dozen small ones at one sitting, with no sauce at all except their own natural, salty tanginess.

When they are a little larger, but still not much over 3 inches in length, Quahogs are excellent steamed. Put at least a dozen per person in a kettle with a cup of white wine, and let them steam for 20 minutes, or until all the shells are open wide. Remove the top shells and serve hot on the half shell with Melted-Butter Sauce and the broth from the steaming-kettle. Good with beer, and excellent with a light, dry white wine.

These middle-sized clams are also just the right size for Clams Casino. Make a sauce of ¼ pound of melted butter, 3 crushed garlic cloves, 1 teaspoon lemon juice, 1 tablespoon finely chopped parsley, and 2 tablespoons chopped chives. Mix all this together and let it stand in a warm place for several hours for the flavors to amalgamate. Open 6 to 12 clams per person, depending on the size of the clams and the appetites of the persons, and remove the top shells. To each clam on the half shell add 1 teaspoon of the Melted-Butter Sauce, then cover with ½ slice of bacon. Place under the broiler, about 5 inches from the unit or flame, and broil until the bacon is just crisply done, then serve hot.

When the Quahog gets above 3¼ inches in diameter, it still has a fine flavor but, let's face it, it's just too tough to eat in any of the above ways. A peculiar thing about this clam is that the further south it goes the tougher it gets. I have eaten 3½-inch Quahogs from Massachusetts that were tender and delicious, while one of the same size from South Carolina or the Sea Isles of Georgia is likely to be tough as sole leather. The big clams are, however, the best kinds for making Clam Chowder and Stuffed Clams.

I like the old-fashioned New England Clam Chowder, and consider it a sacrilege to add tomatoes to this dish. I start with 1 large onion and about 4 slices of bacon, both cut fine and fried together until the onion is golden yellow and translucent. Then I add 2 cups of diced potatoes, cover all with water, and boil until the potatoes are done. While this is cooking I steam 2 dozen clams, or a few more if they are plentiful, grind the meat in a food chopper, then add this ground clam meat and the broth from the steaming-kettle to the chowder. Immediately I pour in 1 quart of milk and add 1 teaspoon monosodium glutamate and ¼ teaspoon freshly ground black pepper. While all this is heating only to a simmer, I blend 1 tablespoon flour in ¼ cup of milk, then slowly stir this into the chowder and continue stirring until it thickens slightly. I try never to let it boil after adding the milk, but allow it to simmer about 10 minutes after adding the thickening, then serve hot with crackers and a tossed watercress salad, and no other food is needed at that meal.

Stuffed Quahogs are frankly fancy and they are usually appreciated by everyone, even those who are not usually seafood addicts. I like a clam-rich stuffing, so I use the meat from at least 2 clams for every half shell I cook. They are not at all hard to make, and this is one of the very finest ways of using those oversize Quahogs you bring home. You will require 2 to 3 large stuffed half shells per person. For four people, steam 2 dozen Quahogs only long enough so they will open wide, then grind the meat in a food chopper. To this add ¼ cup minced onion, ¼ cup chopped celery, ¼ cup chopped green pepper, and 2 cups fresh bread crumbs. By fresh bread crumbs I mean those you crumble yourself from day-old bread, not those bone-dry, ages-old crumbs you buy in shaker packages in the store. Dampen the stuffing with broth from the steaming-kettle and mix well. Pick out 12 uniformly sized half shells, fill them with stuffing, and then lay ½ slice of bacon over each one. Bake in a hot oven, 450°, until the bacon is crisp. Stick a jaunty sprig of parsley in each of the clams and serve them proudly.

The best camp recipe for clams that I know uses this same stuffing. Butter the insides of shells and stuff both sides. Fit the two sides of the shell together and wrap with a double thickness of aluminum foil to hold it together. Roast in hot ashes in the campfire, or cook for about 15 minutes on each side over the charcoal grill. Since these do not have cooking bacon to lard them with savory drippings, they should be served with hot, spicy Melted-Butter Sauce, page 21, poured over them liberally after they are opened. These can be made before your guests arrive and kept in the refrigerator, or they can easily be transported to the picnic site in your portable icebox. Try them on your tony friends; then sit back and watch your reputation as a talented outdoor chef grow.

4. Crabs and Crabbing

The cruelest creature's the crab
With claws that can pinch you or stab,
And then when you dine
On crab and white wine
It's murder when you pick up the tab.

I SELDOM suffer from the kind of "murder" described in the above limerick, for I derive fully as much pleasure from catching my own crabs as I do from eating the many wonderful dishes for which they provide the materials. When I buy crabs or crab meat in the market, or order crab dishes in a restaurant, I always feel that I have been cheated of half the fun. Like those quaint old English recipes for cooking hares, all my crab recipes should begin, "First you catch the crab."

The American coasts are blessed with many kinds of edible crabs, all delicious to eat, smart to serve, and grand fun to catch. Each coastal district seems to have developed its own techniques for crabbing, but I have found that many of the techniques used can be transferred from district to district and from species to species with great success, for most crabs are not at all particular about how they are caught. Crabbing makes a wonderful family sport. Some of the methods are so simple, and some species of crab so easily caught, that the youngest toddler in your family can enjoy the thrill of catching his own crabs with very little help from you. The equipment is simple and inexpensive, the crabs are abundant and delicious, and the roads are good between here and the sea. Let's go.

In this chapter we will first discuss the kinds of crabs you can expect to find along the tidal shores nearest you and the methods of catching them, and then, since crab-meat recipes can be used for almost any species, we will talk about the techniques of cooking and serving this finest of crustaceans.

THE PUGNACIOUS BLUE CRAB
Callinectes sapidus

Commonly called the Blue Crab in the North and the Sea Crab in the South, this is the best-known edible crab of the East Coast. Economically it competes with the lobster and the shrimp as our most important crustacean. From Cape Cod to Florida, these crabs are taken in immense numbers for the market, with the industry concentrated in and around the Chesapeake and Delaware bays. As I live within easy driving distance of both these bays, I meet the Blue Crab many times each summer and fall, and this scrappy crustacean has probably furnished me with more solid fun and delicious food than any others of its kind.

The Blue Crab is at home on muddy shores and is commonly found in great numbers in bays, in tidal streams, and at the mouths of estuaries. An average adult is about 6 inches across the back and is roughly twice as broad as it is long. A distinguishing feature is a long, sharp spine projecting straight outward on each side of the carapace. Between this spine and the eye, on each side, are found 8 short spines, and between the eyes are 4 unequal "teeth" and a small spine beneath.

Its claws are large, and it can use them with great strength and dexterity. Back of the claws are 3 pairs of simple pointed feet, and then the back pair of feet, which are flattened into oarlike swimming members. It is a colorful creature, dark green above, white below, with the feet blue and the tips of the spines red.

Only male crabs can be taken legally in some states, and this is as it should be if we are to preserve an abundance of crabs for future years. Actually, crabs seem to produce a great surplus of males, or else the females are wiser at avoiding getting caught, for among my catches I find that males predominate four or five to one. The sexes

Blue Crab.
*Size: about 6 inches
across back, and about
twice as broad.*

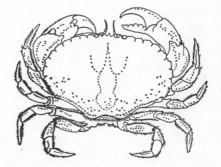

Right, Jonah Crab.
  Size: about 3 inches by 4 inches.

Below, Lady Crab.
  Size: about 3 inches in diameter.

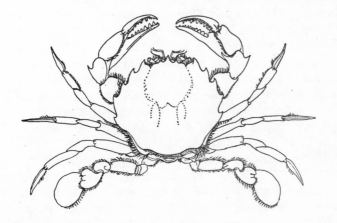

can be distinguished by the shape of the abdomen, usually called the "tail" or "key" by laymen. On the male it is narrow and pointed, and he keeps it neatly tucked under him into a ready-made recess in his plastron or under-shell. On the female the abdomen is broad and short, filling the space between the swimming legs. Usually hers, too, is kept tight against the plastron, but during spawning season it is used as an egg basket, and at this time it is carried almost at right angles to the plastron, with the angle completely filled with eggs.

Those who say that crabbing is tame sport should have been with me one day when a friend and I were out sailing on Chesapeake Bay. It was late afternoon and we had lost our breeze, so I went forward and lay down on the deck looking over the side into the clear, shallow water. The first thing that caught my eye was a large Blue Crab looking up at me, and then I saw another, and still another. We dropped the anchor and went overboard with diving masks, swim fins and a long-handled dip net apiece.

The next hour was filled with hilarious fun. Blue Crabs were on the bottom all about us, and we were down there fighting them in their own element. The Blue Crab is predacious and defiant, and while a few of them made half-hearted attempts to run or swim away when they first sighted us, they would soon turn, throw up their vicious claws, and offer to do battle. Then it was a grand fight getting them into the net and over the side into the boat without allowing them to escape on the way. All this had to be done swimming, for although the water was shallow enough, we couldn't wade without stirring up mud that would hide our quarry.

We kept no count of the crabs we caught, and when we finally climbed back aboard they seemed to be everywhere. There were crabs in our bunks, in the lockers, and underfoot. We finally corralled them into a bushel basket and started heating a pot of water on our tiny galley stove. When it boiled, we dropped in a dozen big ones and prepared to feast royally on crab alone. As we were sitting in the cockpit cracking and eating hot, juicy crab, my friend suddenly screeched and almost jumped overboard. One crab that had been overlooked when we gathered up our catch had climbed up behind the cockpit seat and nipped him in a tender portion of his anatomy.

The abundance of crabs still on the bottom of the bay all around us made us avaricious, so we decided to make our outing pay by catching a bushel or so of crabs to sell. By this time it had grown quite dark, so I shone a flashlight on the bottom and saw a host of crabs making a cannibalistic feast on the refuse we had thrown over the side as we ate. Far from being afraid of the light, they seemed to be attracted by it. By thrusting a net against the bottom and moving the beam across it, we could entice the crabs right into our nets. This was the first time I ever saw crabs help to catch themselves. We kept them coming aboard until the land breeze and the returning tide changed our minds. Next morning we sold two bushels of crabs and still had some nice ones left for our own use.

Fortunately, there are many other ways of taking Blue Crabs. The simplest of these is with a baited line. The equipment required for this method is a long-handled dip net, some stout line, a few fish heads or other bait, and some lead sinkers.

When I clean fish, I drop the heads into a plastic bag and store them in my freezer, so I always have a supply of crab bait on hand. There is a widespread belief that crabs prefer ripe, smelly baits, but the few times I've had the opportunity to try fresh and smelly baits side by side I have always taken more crabs on the fresher, sweeter-smelling baits. Certainly fresh or frozen baits are far more pleasant to handle than half-rotten fish, and one feels better about eating the crabs caught on it.

Crabbing with baited lines can be done from the shores of tidal streams or from docks, bridges, or boats. Tie a fish head on your line, weight it with a lead sinker, and rest it on the bottom. One crabber can handle a dozen lines. Once you're thoroughly practiced, you can usually tell whether a crab is on the bait merely by lifting the line until the sinker clears the bottom. Lift the crab smoothly to the surface, slide your net under it, and lift it out. That's all there is to it, and when you are in a good spot and conditions are just right, you can sometimes get a boatload of crabs on a single tide using this method.

When I go fishing for perch, drum, and Striped Bass in bays and tidal streams, I sometimes catch enough crabs for dinner on whatever fish bait I happen to be using. When more crabs than fish are coming aboard, I never groan or swear as some fishermen do. I am grateful for any good gift from the sea, whether it is the one I asked for or not.

The two methods that offer the most promise of success for the neophyte crabber are the folding crab trap and the ring net. I mention these two together because they are used in exactly the same manner. The folding traps are of two kinds, one with a fixed bottom and top and four sides that let down when the trap is rested on the bottom, and another kind with only a fixed, triangular bottom and three sides that fold up into a three-sided pyramid. I seem to have better luck with the three-sided ones, though I could never explain why. These traps would be difficult to make at home, and as they are reasonably priced I recommend that they be bought ready-made. You can get three of them for less than five dollars, and in good crabbing territory that many will practically put you in business. Ring nets are wire barrel hoops with fish netting laced inside them and a bridle and lifting string attached. Ring nets are standard crabbing equipment on the West Coast and in Hawaii, but are seldom used in the Blue Crab country. There seems to be no reason for this neglect of a potent method of catching crabs except custom. The Blue Crab certainly has no prejudice against being caught in this manner, for I have pulled up as many as six at one time on a ring net.

In using either the folding trap or the ring net, tie your bait firmly to the center of the base, lower the device to the bottom for a few minutes, then pull it up. Any child able to lift a trap or net to the surface can do as well at this kind of crabbing as anyone can, so the whole family can get into the act.

Finally, there is the set trap used by commercial crabbers. These are no more than lobster pots with the shape of the entrance changed to fit the shape of the crab. In use they should be baited, and they work better in open bays in comparatively deep water than they do in muddy tidal streams where mud snails are apt to eat the bait before the crabs find it. This kind of trap is usually attended only once a day.

SOFT-SHELLED CRABS

The Soft-Shelled Crab, beloved by gourmets, is not a separate species as some think, but is our old friend, the Blue Crab, just after one of its periodic sheddings of its hard shell. The hard chitin covering of the crab is an excellent piece of equipment, serving as both armor and skeleton, but it does hamper growth. The only way a small crab can grow into a large one is by periodically molting each shell as it outgrows it and then growing a new one to fit its larger size.

When the crab sheds its shell, it does a thorough job, discarding the outer covering of its feet, eyestalks, and even its antennae, as well as the hard body shell. For a few days it is soft, white, and pretty helpless, not at all like its usual scrappy self. The Soft-Shelled Crab is a great delicacy, and you can eat the whole thing, the outer covering as well as the delicious meat within. The soft white shell of the crab at this stage not only tastes good, but it contains some excellent proteins and essential minerals that help to keep us healthy and vigorous.

Where I usually go crabbing, Soft-Shells are in season during June, July, and August, but as only a small percentage of the crabs in any one area will be molting at the same time, don't expect the bottom to be covered with Softies. Sometimes, when the weather is bright and calm and the water is clear and shallow, one can see the lethargic Softies crawling sluggishly over the bottom, and in this condition they can be picked up by hand with no danger of getting pinched. A diving mask or a glass-bottomed box will be a great help in locating them. On the whole, however, trying to gather a supply of Soft-Shells from the wild is a very haphazard and uncertain business, seldom attended with success.

The Soft-Shelled Crabs one sees on the market were captured as Hard-Shells and held captive until they molted. A few years ago, I started using a heavy wire-mesh trash burner as my own private crab pen. At first I kept this amateur Soft-Shell factory a deep secret for fear of public and family ridicule, but when it began producing more Softies than my family could eat, I decided it was safe to reveal my methods.

In use the pen must be moored in a tidal current where there is a constant circulation of water through it. This is very important. If crabs are confined in still water, they soon use up all the oxygen about them and promptly drown. The second important point in the successful operation of a pen is selecting the right crabs to be confined. If one simply dumped each day's crab catch in the shedding pen and waited for all of them to molt, one would have a long wait. Those about to molt are called "peelers" or "shedders" by commercial operators, and to distinguish these candidates for the shedding pen you look at the jointure of the carapace and the plastron between the two back swimming legs. If the back seam between the top and bottom shells is neat and tight, you have a "green" crab that won't molt for a long time to come. However, if the two shells are beginning to part, and are only joined by a stretched semitransparent membrane

through which the new shell shows as a white line, you have a shedder
that will soon produce a delicious Softie.

When I am vacationing at my favorite locality on Delaware Bay,
I catch crabs every day, either purposely or as a pleasant by-product
of my fishing. I look each catch over, imprison the shedders in my
homemade pen, and eat the rest as Hard-Shells. With a pair of
kitchen tongs you can handle the crabs without risking a painful
pinch.

ROCK CRABS AND JONAH CRABS
Cancer irroratus and *C. borealis*

These are the best-known edible crabs of the rocky New England
shores. The Rock Crab ranges from Labrador to Cape Hatteras, but
is rare south of Long Island. It is sub-oval in outline and broader than
long; a good-sized specimen will measure about 4 by 3 inches. The
back is yellowish in color, closely dotted with brown, and the surface
looks granular but feels smooth. There are 9 blunt teeth along each
side of the forward edge of the carapace. All species of the genus
Cancer are walking crabs, that is, the rear pair of legs terminate in
pointed feet rather than being broadened into oarlike swimming
members.

Unlike the more southern Blues, Rock Crabs prefer rocky or sandy
bottoms to muddy situations. They are found in shallow water just
below low-tide line, or sometimes even hiding among the rocks between the tide lines.

Natives of New England usually do not distinguish between the
Jonah Crab, *Cancer borealis*, and the Rock Crab described above,
but there are some differences between these two closely allied species.
The Jonah is larger and more massive than its kinsman, and the legs
are proportionately shorter and heavier. The shell is brick-red above
and yellowish beneath, and the claw and carapace are much rougher
than those of the Rock Crab, the granules being irregular in size. It
too has 9 teeth along the anterior margin of each side of its upper
shell, but on the Jonah the forward teeth are rounded, while the after
2 or 3 on each side are sharply pointed. Jonahs are rather local in
distribution, being abundant only in some areas from Nova Scotia to
the eastern end of Long Island. They are always found on rocky shores
not far from the open sea.

For some strange reason that I will never understand, New Englanders seem to prefer to leave crabbing to the professionals. If you are a resident of New England, you can obtain a license to catch crabs with crab pots in your own state, but this method is forbidden by law to the visitor or tourist. However, I have proved to my own satisfaction that Rocks and Jonahs can be caught legally by amateur methods as easily as any crabs anywhere. On a recent trip along the coasts of Connecticut, Massachusetts, New Hampshire, and Maine, I was able to catch all the crabs I wanted, and these New England crabs are so good that I wanted a great many of them. I first tried the folding crab trap and the ring nets, and if there were any crabs in the locality, I caught them. The tremendous tides of northern Maine make it difficult to crab from wharves or piers with baited lines, but from a floating dock or boat this method also worked well. Both the Rocks and the Jonahs seized the bait as readily as the southern Blues and were easily brought up within netting distance. While poking around the shore and tide pools at extremely low tide, I discovered the easiest way of all of catching new England crabs. This was simply to scoop them from the bottom with a dip net. Every time I tried it, I got all the crabs I could use during the hour or so of lowest water. If you find New England seawater unbearably cold, a pair of fishing waders makes this kind of crabbing much more comfortable. Look around rocks, in weedy pools, and especially where the bottom is grassy.

Whenever I went crabbing in these Northern states, some of the local boys would invariably flock around and marvel at these easy ways of catching crabs, so I may have introduced this sport into a new area.

My wife and I are both very fond of crab meat, but she would rather not see the crabs while they're still in the shell. Each day when the tide practically bared the bottom, I would go out and capture a quarter bushel of crabs, sneak into the kitchen of our cottage and boil them for 10 minutes, then take them to a shady place and extract the meat with nutcracker and nutpick. These crabs have a surprising amount of meat in the first joints of their large walking legs, and on very large Jonahs, even the second joints are worth opening. I would take the meat back to the kitchen, add a bit of salt and 1 slightly beaten egg, and shape it into little balls with damp hands. The kind of crabbing my wife likes to do is to return from an afternoon's swim or sunbathing session, open the refrigerator door, and find a generous plate of Crab Cakes all ready to cook. We cooked them in a variety of ways. They were good just baked on a cookie sheet or fried in deep or shallow fat. Once we made a Crab Gumbo with frozen okra (page 46) and instead of merely adding loose crab meat we dropped small Crab Cakes into it and boiled it for an extra 6 minutes. Served on a helping of wild rice, this made a dinner that could not be soon forgotten.

THE SOUTHERN STONE CRAB
Menippe mercenaria

This species ranges from North Carolina around the Florida peninsula and the Gulf Coast to Texas. Stone Crabs live in deep holes in the mud along the shores of tidal creeks and estuaries, in rock heaps or drifts of debris. Adults are a little wider than long, oval in outline; a good-sized one will measure about 3 by 4½ inches. One claw is usually larger than the other, but both are proportionately huge and tipped with black. The terminal joints of the other 4 pairs of feet are thickly fringed with hairs and end in points like spikes.

I once had the pleasure of hunting Stone Crabs with a bright Negro lad who was financing his education by crabbing after school and on weekends and selling his catch directly to fancy restaurants. I should say I had the pleasure of watching him at his work, for I hadn't the courage to catch crabs as he did. At low tide we waded along a muddy inlet where burrows of the Stone Crab abounded. My companion would kneel in the mud before each occupied burrow and thrush his whole hand and arm into it, fold up the vicious claws of the crab within, then grasp the creature tightly, jerk it from the hole, and drop it into a string bag which was his only equipment. He seemed to have eyes in the ends of his fingers, for he was never pinched. It looked easy, but when I tried to emulate his example and follow his instructions, I was badly pinched on my first and only try. He then admitted that he had suffered a few mangled fingers while serving his apprenticeship at this unusual trade.

Later I developed a less painful method of removing Stone Crabs from their burrows. Use a pair of kitchen tongs, and thrust them into the hole with the jaws wide open. When you feel the hinge, where the two handles of the tongs join, strike the crab, clamp down hard and instantly pull the crab all the way out of the hole in one motion.

If you hesitate a moment, the Stone Crab will brace itself so tightly against the sides of its burrow that you will find it very difficult to remove.

The back-fin or body meat of a Stone Crab is boiled, picked out, and prepared like that of other crabs, but the tremendous claws, rivaling those of a lobster, should be served separately. These should be boiled for 10 minutes, then served either cracked and in the shell, or picked out without breaking the meat and served with Melted-Butter Sauce. These huge Stone Crab claws, without the crab attached, have recently made their appearance in Northern markets, selling for dollars per pound. Rich epicureans meekly pay the outrageous prices for these big claws, for there is no finer seafood in existence, except those that have that extra magic that only comes when you have gathered the food with your own hands, fresh from Southern tidal streams.

THE LADY CRAB, SAND CRAB, OR CALICO CRAB
Ovalipes ocellatus

This is a beautifully colored and patterned crab found on sandy shores from Cape Cod to Florida. It is common among the loose sands at low-water mark of even the most exposed beaches and is often very abundant on sandy bottoms offshore. Only about 3 inches in diameter, this crab is about as long as it is wide. There are 5 prominent, pointed spines on each side of the forward edge of the carapace, then the indentations for the eyes, and between the eyes is a 3-spined rostrum. The coloring is gay, being a ground of white covered with spotted rings of red and purple. The first pair of legs are large and equipped with toothed claws capable of giving a painful nip; the posterior pair of legs are flattened on their tips into oarlike swimming members, and the 3 intermediate pairs are simple in structure, ending in sharp points. The common name, Lady Crab, must have been given because of its bright dress and not its behavior, for this is the crab that most frequently nips the feet and legs of bathers off the sandy beaches of East Coast resorts.

Like other animals that inhabit exposed, sandy beaches, the Lady Crab can burrow. It often buries itself to the eyes just at low-water mark. I seldom go surf clamming without getting a few painful pinches from the Lady Crabs I inadvertently come across while feeling about in the sand for Surf Clams. If I come across enough of them, I take revenge by first boiling and then eating them.

The Lady Crab is edible, even delicious, despite its small size, which makes picking out the meat somewhat tedious. Large numbers can be caught in shallow water just beyond the surf of exposed beaches by using ring nets, folding traps, or small-meshed crab pots. The right bait for all three is clams. On calm days I have caught dozens of Lady Crabs by wading out to waist-deep water and setting my ring nets baited with the tough feet and viscera of the Surf Clams I was taking from the beach. My favorite method of cooking Lady Crabs is to boil them in seawater over a campfire at the beach, then eat them while they're still hot. I'll admit that it is a tedious job to pick the meat from a Lady Crab, but can anyone tell me a more pleasant way of passing a long evening around a campfire?

THE BATTLING GREEN CRAB
Carcinides maenas

The Green Crab ranges from southern Maine to New Jersey, but is most common between the tide lines from Cape Cod to Long Island. In size, shape, and number of teeth on the forward edge of the shell it resembles the Lady Crab, but the Green Crab has a colorful green carapace spotted with yellow, making it very conspicuous. Although it is a member of the swimming-crab family, its hind legs are not modified into swimming members nearly so much as are those of the Ladies and Blues. On the Green Crab the back pair of legs are truly combination limbs, being slightly flattened for swimming but retaining their points for walking also.

This species is so abundant in some sections that every loose stone will have from one to a dozen crabs under it at low tide. When you first turn over a loose stone near the water's edge you may think there is nothing there, for the Greenies bury themselves just under the sand. Rake your fingers through the wet sand under the stone, and you will often find that you have half a dozen belligerent little crabs all threatening to do battle to the death. By searching carefully you may also find a number of lethargic Soft-Shells that are hiding in the sand under the rocks after recently shedding their armor and waiting to grow a harder shell before venturing forth again.

When I speak of eating Green Crabs, most people, even those who live where this crab is abundant, will say that they never heard that it was edible. All true crabs are edible; it's just a question of how much work you are willing to do to get the meat from the smaller species. I find picking the meat from the Green Crab little more tedious than from the Rocks and Jonahs; the Soft-Shells are delicious fried, broiled, or sautéed, and you can eat the whole crab when it is in this stage.

The Green Crab is also found along the Atlantic shores of England and France, and the seafood-loving French have also learned to appreciate its culinary possibilities. They must also have noticed the reckless courage with which it will attack a human being many times its size, for they call this game little scrapper *crabe enragé*.

THE WEST COAST CRABS

By far the most important crab on the West Coast is *Cancer magister*, a close cousin of the New England Rock Crabs and Jonah Crabs. This is the Dungeness Crab, famous among epicures and found on the menus of fine restaurants all over the nation. It is also commonly called the Edible Crab, or the Big Crab, and it does grow to be a whopper. Specimens have been taken that measured 12 inches across the carapace, but this is most unusual. In the spots where I've gone crabbing, most individuals were from 6½ to 7 inches across, and one that measured 8 inches would be something to write home about.

The carapace is finely granulated, of a reddish-brown color, and there are 10 teeth on either side of the anterior margin, the one farthest aft being decidedly the largest and quite sharp. These 2 large teeth are the "horns" or "points," and one measures between them to determine whether or not the crab is of legal size.

The Dungeness, although named for a locality in the state of Washington, is found all the way from the Aleutian Archipelago to Baja California. It is taken very readily with ring nets or folding traps about the mouths of bays, on tidal flats, and in most inland tide waters. Commercially the Dungeness is taken in crab pots baited with fish heads and other offal. It is found in both sandy and muddy situations, but prefers sandy bottoms.

One morning I chanced on an exceptionally low tide, a few miles north of Seattle, Washington, about sun-up. I took a shovel and a pail and started out to dig a mess of clams for dinner. The best nearby clamming beach was across a shallow bay from my cottage. Rather than walk all the way around the head of the bay, I decided to wade straight across, for the water was nowhere more than three feet deep with so low a tide. As I was wading along in knee-deep water, I saw a big crab scuttling along sidewise ahead of me. I speeded up in pursuit, and he soon turned and threw up his big claws in the characteristic manner of a crab at bay. I put my shovel down before him, and he very obligingly climbed onto it and soon found himself in my pail. Once alerted I soon saw other crabs, and in the excitement of the chase I forgot about clams. In less than an hour, my pail was so full of crabs that they were beginning to escape over the top. I still had time before the tide came in to fill my pockets with enough Butter Clams to make a delicious chowder to start off a wonderful seafood dinner that featured Crab Louis as the main course.

Two other Pacific crabs that deserve more than casual mention are *Cancer productus*, the Red Crab or Sea Crab, and *Cancer antennarius*, a Rock Crab. Both these excellent edible crabs are found near rocky shores from Kodiak, Alaska, to Lower California. They are more abundant than is generally known; they seldom appear in the markets, not because of their rarity but because of the difficulty of catching them in commercial quantities on the rocky shores they inhabit. Both these species are sometimes taken in ring nets, folding traps, and crab pots about rocky inlets, but the catch is not as dependable as is the case with the Dungeness. On calm days these crabs can often be taken with a dip net from rocky bottoms at extreme low tide, using a diving mask or a glass-bottomed box to help in seeing them. Unlike the Dungeness, these two crabs sometimes invade the intertidal zone, and at night they can be hunted in the tide pools, where you will find them strutting about, lording it over all the smaller life-forms. On an extremely low tide I have found Red Crabs hiding in the sand around the bases of rock piles.

This Rock Crab, *antennarius*, is dark purplish brown on the back and red-splotched underneath. The teeth along the anterior margin of the shell are large and distinctively hooked or curved forward. You will seldom catch such Rock Crabs more than 5 inches across the carapace, but because of their proportionately enormous claws one specimen this size will yield as much meat as a considerably larger Dungeness.

Cancer productus is a larger species, rivalling the Dungeness in size. The adults are dark red above, appearing almost as if they had already been boiled. The carapace juts forward between the eyes, and this portion of the shell is made up of 5 nearly equal-sized teeth, evenly spaced. Some epicures maintain that this is the finest crab on the West Coast, even excelling the Dungeness in flavor. The meat is so delicious that I recommend that this crab be cooked by boiling in plain seawater and the meat be eaten directly as it is picked from the shell, with no seasoning or other ingredients to dilute that wonderful flavor.

Southern California and the Pacific side of Lower California are poor in really fine crabs. This unfortunate distribution of edible crustaceans has led to the widely prevalent belief that really good crabs are only found on the shores of northern seas. This is nonsense. Our newest state, Hawaii, with all the main islands lying wholly within the tropics, has four different species of edible crabs in its surrounding waters. One of these, the Samoan Crab, *Scylla serrata*, grows larger than any crab commonly caught around mainland shores, and both this one and another, the Kona Crab, *Ranina serrata*, are unsurpassed for flavor. I should explain to nontechnical readers that similarity, or even identity, in specific names does not denote kinship, and the two Hawaiian crabs, both bearing the specific name of *serrata*, are vastly different from one another in both appearance and habits. When you take a vacation in Hawaii, don't miss the excellent crabbing to be found there, and some evening while in an expansive mood, dine on Samoan or Kona Crab.

While Southern California and western Lower California are poor in crabs, they are not absolutely destitute. All three of the large crabs of the genus *Cancer* are occasionally taken in southern waters, and there is a relative of our Eastern Blue Crab that is native to these parts. This native, *Portunus xantusii*, commonly called the Swimming Crab, would make an excellent dish if it would only grow a little larger. As it is, they are not at all bad if one has enough of them, even though a large one is only about 3 inches across the carapace. Any real crab fancier will be willing to pick the meat from this delicious pigmy crab despite the tediousness of the job.

This Swimming Crab is easily recognized because it is the only true Swimming Crab of this size in these parts, that is, it is the only one with the hind pair of legs flattened into oarlike swimming paddles. The front pair of legs are proportionately enormous, making up about half the weight of the entire crab, and they terminate in a pair of sharp claws that are handled with great skill and effect. Inside these big legs and claws are found the largest pieces of meat you will be able to locate in this crab. The little Swimmer is the most lively and active of all crabs, and it uses those strong claws with great speed and dexterity at the slightest opportunity. Anyone who catches and confines enough of these crabs for a meal without contributing a few drops of blood must be very quick and very careful.

I once read a statement that *Portunus xantusii* was never found in the intertidal zone except around sand flats that were regularly covered and uncovered by the tides, and it is perfectly true that channels leading to such sand flats furnish some of the very finest crabbing spots when you are seeking this species. However, I have caught dozens of these little crabs from San Diego Bay where samples of the bottom looked very much like ordinary mud. Fish for them with ring nets, folding traps, and crab pots, and they can be caught, in suitable situations, from San Pedro to South America.

I have never heard of anyone confining "shedders" of this species to produce Soft-Shelled Crabs, but I'll bet it could be done. Since there is so much more to eat with a Soft-Shell, this would remove some of the objection to this crab's small size. Four or five Soft-Shells only 3 inches across the carapace would make an ample serving for even the best of trenchermen. Why don't some of you West Coast crabbers try it?

Baja California boasts of one other truly excellent crab, the Gulf Blue, this one found along the Gulf of California shores of this interesting peninsula. This is a much larger crab than the one just discussed, another close relative of our Eastern Blue Crab, indeed in the same genus, for biologists have classified it as *Callinectes bellicosus* and it certainly deserves that specific name, for you will seldom meet a more belligerent fighter. I was unable to obtain a specimen when our artist was making the illustrations, but you will have no trouble recognizing this species, because it is the only Swimming Crab of what is considered *edible* size along this coast. As large as the Eastern Blue and with the same bright-blue claws and legs, this crab differs from its Eastern relative in the shape of its body, being

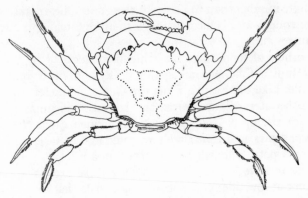

Above, Green Crab. *Size: about 3 inches in diameter.*
Below, Dungeness Crab. *Size: about 6½ inches across.*

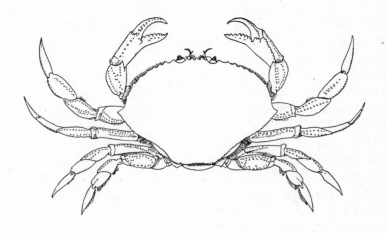

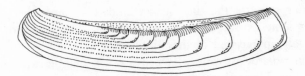

Atlantic Razor Clam. *Size: about 6 to 7 inches long, 1 inch across.*

much longer in proportion to its width than is its Chesapeake cousin. It is large enough and bellicose enough to make catching it anything but a tame sport.

As an article of food, this crab is highly appreciated by the Mexican and Indian fishermen along this coast, and any surplus finds a ready sale. Like the Eastern Blue, the Gulf Blue prefers to stay in the brackish waters of bays, inlets, and lagoons, and shuns the salty brininess of the open Gulf. The local fishermen catch them with ring nets or seines, and I presume they could also be taken in folding traps or crab pots, although I have never tried it.

The way to have the most fun catching these aggressive crabs is to wade out at low tide on the shallow sand flats that make the bottoms of the many lagoons to be found along this coast and chase them until they come to bay, then scoop them up in a dip net. They are often abundant enough in such localities so that one chase succeeds another as fast as you can scoop them up, until you are either too tired to continue or else have all the crabs you can use. Like other Swimming Crabs, these Gulf Blues have their hindmost legs modified into very efficient swimming paddles that can propel them through the water quite rapidly. However, this is not exactly a timid creature and it seldom swims far before it turns on you and offers to do battle.

Gulf Blues are delicious if cooked soon after being caught, but one should never try to keep them uncooked for long, nor try to transport them far, for they soon die when taken from the water. This is another species that almost certainly would furnish some excellent Soft-Shelled Crabs if one practiced on them the same techniques used in denuding our Eastern Blues.

I cannot leave the subject of the edible kinds of crabs without mentioning some of the tiny, tender kinds that share the bed and board of some of our finest edible mollusks. The best-known and most highly appreciated of these interesting creatures is the Oyster Crab, *Pinnotheres ostreum*, which lives in the mantle cavity of ordinary Oysters. These aberrant crustaceans take up their strange abode while still in the larval stage, and they may grow to be 1 inch across. Their relationship with the Oyster is neither parasitic nor symbiotic, that is, they neither hurt nor help their host, but merely live there and share the food pumped in by the Oyster's siphons. Naturalists call this a *commensal* relationship.

Strangely enough among these commensal crabs it is only the females that lead such a secluded life. While we think of these females as small animals, the males are many times smaller, and they are free-living, only popping into an Oyster shell now and then to fertilize any available eggs, thus insuring future generations of Oyster Crabs. The easy life has caused the females to degenerate; their eyes have become useless, and their shells never harden, making them permanent Soft-Shelled Crabs. Their color is pale pink, and they are very slow and sluggish in their movements.

Since ancient days, these tiny soft crabs have been highly esteemed by discriminating epicures. I have a friend who operates an oyster-shucking shed, and since the average, unknowing diner doesn't like to find crabs in his Oysters, my friend keeps these delicacies for himself. Knowing how I appreciate such fine fare, he sometimes presents me with a pint or so of these dainty pinkish Soft-Shells. They need no cleaning whatever. One can sauté them in butter until they barely begin to turn golden brown, then serve them on toast with tartar sauce on the side. They need no seasoning, for they are naturally salted to perfection. If you prefer them fried, dump the whole lot into a mixing bowl containing 1 egg beaten with 2 tablespoons cold water. Stir them around, then drain off the surplus egg and shake them in a paper bag with 1 cup of fine cracker crumbs. Fry until lightly browned in bland oil heated to 375°. Each little crab makes but a single bite, and cooked either way they are some of the most delectable food that ever came to the table.

The Blue Mussel also has a commensal crab, called the Mussel Crab, sharing his shell and food. This is *Pinnotheres maculatus*, and when I find these little crabs in Mussels I cook them with their hosts and eat them both together.

Along the Pacific Coast the huge Horse Clam, *Schizothaerus nuttallii*, plays host to a commensal crab that rivals the Oyster Crab in flavor. This is *Pinnaxa faba*, commonly called the Pea Crab, which grows to about 1 inch in diameter. In some parts of Puget Sound I found every Horse Clam had at least one Pea Crab living inside, many had two, and some had as many as three. When these were breaded and fried, they were almost as good as Oyster Crabs, and that is good enough for any food to be.

CREATIVE CRAB COOKERY
CLEANING SOFT-SHELLS

You will need at least one large or two small Soft-Shells for each person to be served. Cleaning a Soft-Shell is easy, for in this condition it is far from being its usual belligerent self and can be handled with impunity. With a sharp knife remove the eyes and the stomach, which is the soft substance just below and behind the eyes. Make a slit along each side, fold back the top skin a bit, and remove the "devil's fingers," those spongy strips just under the back. Rinse the crab in cold running water and the job is done.

FRIED SOFT-SHELLS

Beat 1 egg with 2 tablespoons water and with this dampen the crab all over, using a pastry brush. Roll the egg-covered crabs in fine cracker meal, then fry in hot fat, 375°, until they are lightly browned, about 3 to 5 minutes.

BROILED SOFT-SHELLS

Crabs to be broiled require no egg-and-crumb coating. Arrange the Softies in a pan, top each one with a generous pat of butter, and place about 3 inches below the broiler flame or unit. Broil about 4 minutes on each side, spooning the melted butter over the crabs when you turn them.

SAUTÉED SOFT-SHELLS

Melt ½ cup of butter over medium heat in a pan equipped with a tight cover. Put in the crabs, cover, and sauté for about 10 minutes, shaking or turning them occasionally so they brown to an even golden color all over.

Cooked in any of these ways, Soft-Shells should be served with hot toast or crusty French bread that has been reheated in the oven. Garnish the crab with lemon slices, parsley, and watercress, and by all means serve them with Beach-Plum Jelly if you can possibly make it or get it.

Although Soft-Shells are easy to clean and simple to cook, I somehow never think of them as a camp dish. I suppose it is cultural conditioning, but the idea of serving such luxuries on paper plates to people in camp togs or bathing suits seems inappropriate. They are good enough to deserve the splendor of snowy linen, gleaming silver, delicate china, sparkling glassware and the finest wines. I feel faintly embarrassed approaching a well-cooked Soft-Shell in anything less than my best.

HARD-SHELLS

Whether you are going to serve plain boiled crab and let each diner pick out his own, or whether you are going to create the fanciest of crab recipes, the first step is to boil and clean the crabs. When at the beach I boil crabs in plain seawater, but at home I add a handful of salt and ¼ cup of vinegar to the boiling solution. The crabs are simply dropped into the boiling water alive. This may sound cruel, but really it is the quickest and most humane way to kill a crab, as you will discover if you ever try to do it any other way. Allow the crabs to cook 10 minutes after the water comes to a boil again. Dump them into a sink and cool with running water, but don't let them soak. Remove the claws and legs, and crack them with a nutcracker or pair of pliers. Lift the "key" or "tail" from its recess in the plastron and tear it completely away. Now grasp the top shell at the rear end and lift it off. If you intend to make Deviled Crab, save the shell. Remove the spongy "devil's fingers" found under the top shell and the soft stomach under the eyes. Break the body into two pieces along the seam made by the recess for the "key." Wash each half, and you are ready to pick out the meat for any number of delicious crab recipes. A nutpick is the proper tool for lifting the crab meat from its chitinous recesses.

A CRAB-BOIL DINNER

If you intend to give a Crab Boil for your family or friends, cook the crabs exactly as described above, except that you add a bag of crab-boil spice to the boiling water. Crab-boil spice is a mixture of spices, and several companies put it up loose or in bags ready to use. Unlike the delicate and fancy broiled, baked, or sauteéd Soft-Shells, I consider plain boiled Hard-Shell Crab eminently suited to serving on the beach or in camp, preferably with all diners in bathing suits. When serving a Crab Boil at home, avoid all attempts at toniness, for the dish is not compatible with swank. Set up the table in the kitchen, patio, or recreation room where the floor is easily cleaned afterward. Use a plastic tablecloth and furnish plenty of paper napkins. If your guests are dressed, tie a lobster bib about each one at the beginning of the meal. Have several containers of Melted-Butter Sauce on candle trivets so it will be kept hot, and use it often and generously. If you serve potatoes, make it French fries on a side dish, leaving the dinner plate solely for the main operation. Heap salad plates with large sprigs of watercress liberally sprinkled with a bland dressing made of 3 tablespoons cider vinegar, 1 teaspoon monosodium glutamate, and 1 cup pure, cold-pressed olive oil, combined in that order and well shaken. Serve any light, dry, white wine you happen to like. Provide a large container for the emptied claws and shells, so the diners can clear their plates for a second helping. You will need at least 2 large crabs for each person, with a few extra for really accomplished crab aficionados.

Relaxed enjoyment of a Crab Boil is a messy, drippy, finger-smearing business, so let's face it and not demand any special table manners for this occasion. Encourage each diner to extract those moist, warm, delicious bits of meat in the most efficient manner he can devise, even if it means sucking it from the shell. I know all this sounds as if I were describing a pig's dinner, and there *is* more than a little resemblance, but partakers of a Crab Boil should understand that they are there to enjoy some wonderful food and not to display their clothes, breeding, and table manners.

DEVILED CRAB DELUXE

Pick out the crab meat and wash the empty shells. To 2 cups of crab meat add 1 cup bread crumbs, 1 egg slightly beaten, ¼ cup each of minced celery and green pepper, 1 tablespoon Worcestershire sauce, a sprinkle of freshly ground black pepper, a dash of Tabasco, and 1 cup of unwhipped whipping cream. Mix all ingredients thoroughly and stuff into 4 crab shells, for this recipe only serves four, but it can be doubled or trebled for larger groups. Decorate each serving with 1 or 2 cracked crab claws to advertise the main ingredient, then bake in a 375° oven for about 12 minutes, when the top should be just beginning to turn brown. Garnish with lemon wedges and parsley sprigs, serve while still sizzling, and bask in the well-deserved praise that will be coming your way.

VARIATIONS ON A THEME

One might think that a dish as delicious as perfectly prepared Deviled Crab could hardly be improved, but it can. In the Chesapeake Bay area and southward, the big Blues are still running when the hickory nuts begin to ripen and fall. One autumn, as I was preparing a dinner in which I was trying to use as many natural, foraged foods as possible, I substituted 1 cup of chopped hickory nuts for the bread crumbs of the above recipe, with wonderful results. Since then I have tried using walnuts and pecans when hickory nuts were not available, and in every case I have found crab and nuts have an almost mystical valency for one another; the result is always a very special seafood treat.

On another occasion, my wife and I had promised another couple a dinner of homemade Deviled Crab, and then the crabs failed to cooperate. We had all the excuses fishermen conventionally use: the day was windy, the tide was wrong, and the crabs weren't hungry. We did catch 4 good-sized crabs, but 4 crabs simply will not make Deviled Crab for four people. If you use less than the meat from 3 crabs to stuff each shell, you should, in all honesty, label the dish Deviled Bread Crumbs. Like many another unsuccessful fisherman, I pulled my hat down over my eyes, took a furtive look around and ducked into a fish market, only to find they had no crabs or crab meat. They did, however, have some excellent shrimp, so I bought a pound. On arriving home I shelled, deveined, and boiled the shrimp for just 6 minutes, then diced and mixed it with the pitifully small amount of crab meat yielded by the 4 crabs. I then proceeded according to the recipe for Deviled Crab, and through some quirk of culinary chemistry, the result was better than it would have been had I used either crab or shrimp alone.

Another time two unexpected guests arrived, all the way from England, just as I was beginning to prepare a seafood dinner. I quickly scratched Deviled Crab from the menu and replaced it with *Crabe au gratin*, then dumped one can of condensed cheddar cheese soup and another cup of bread crumbs into the Deviled Crab mixture. I then retrieved 2 more crab shells, and the servings were ample and the dish was good.

There are people, and these include my wife, who claim that the other ingredients and the baking tend to mask the crab flavor in

Deviled Crab. In an attempt to satisfy these people, I devised what I call my Cold Devils. I suspect that one reason the crab flavor is more pronounced in these is that they contain considerably more crab meat, so I only make them when crabs are abundant. To make four servings you will need a full pound of cooked crab meat. To this add 4 tablespoons cider vinegar, 2 tablespoons olive oil, 2 hard-cooked eggs, chopped, ½ teaspoon salt, a little freshly ground black pepper, and just a dash of cayenne. Mix well and pack into 4 crab shells. Spread a little mayonnaise on top, decorate with little trails of paprika, and sprinkle liberally with finely chopped parsley and blanched celery leaves from the center of the stalk. Chill for several hours so it can be served very cold.

SOUTHERN CRAB GUMBO

When I was in the South, I was once served a Crab Gumbo in which the crabs had been boiled, cleaned, and the body pieces broken in halves, the claws cracked, and then cooked with okra, tomatoes, and seasonings. It was served over large plates of rice, and was even messier than a Crab Boil, for one had to pick up the gooey pieces to extract the meat from them. The taste was simply wonderful, but eating it was a very undignified procedure.

I think I understand the insistence of the French, and our own Southern Creoles, that the cook include the shells of all shellfish he puts into the dish. Allow any other procedure and substitutes, fillers, and cheaper ingredients start creeping in. If you don't believe this, try to find out how much crab meat is included in the so-called Crab Cakes sold in most roadside diners. However, if you are making the dish at home you *know* how much of everything you are putting in, and if you try to deceive yourself and your family you will get the poor dish you deserve.

After my initial introduction to Crab Gumbo, I tried refining the dish with considerable success. You will need 16 to 20 crabs to serve six people. Cook and clean the crabs and pick out the meat. Chop 1 large onion and 1 green pepper, and sauté them in 2 tablespoons melted butter until golden. Add 1 tablespoon flour and stir until it browns slightly, then add 3 cups strained, cooked, or canned tomatoes and 2 cups sliced okra. You can use 1 package frozen okra. or 1 one-pound can of canned okra, if the fresh product is not available in your area. Season with 1 teaspoon celery salt, ¼ teaspoon black pepper, 1 teaspoon sugar, 3 or 4 dried Bayberry leaves, and 1 tablespoon gumbo filé.

Gumbo filé can be bought at fine-food and grocery stores, but you can easily make your own supply. It is nothing more than sassafras leaves dried at room temperature, then crushed and sifted to remove the larger stems and veins. Small quantities added to almost any soup or stew will improve both the flavor and the texture of the dish.*

Barely simmer the gumbo until the okra is tender, about 15 minutes. Stir in the crab meat and 1 teaspoon monosodium glutamate, and as soon as all becomes really hot it is ready to serve. The way to eat Crab Gumbo is first to make a bed of fluffy rice on your plate, then ladle the gumbo over it. If you want to achieve the absolute ultimate, use wild rice. I do, because each autumn I gather a supply of this epicurean cereal and do not have to consider those dollars-per-pound prices asked by the grocers. You could probably do the same in your area if you really wanted to ride this hobby.*

CRAB CAKES

Ordinary Crab Cakes are pretty plebeian fare, but they can be excellent if one is generous with the crab meat and severely limits or completely eliminates the other ingredients. At one time I added such things as chopped parsley, minced onion, chopped peppers, chopped celery, bread crumbs, and other things to my Crab Cakes, but by experimenting I gradually made the amazing discovery that Crab Cakes are much better when made of crab. Now I add 1 beaten egg, used for its mechanical qualities, to 4 cups of crab meat and mix thoroughly, then shape it into little 1¼-inch balls with dampened hands. I make these several hours ahead and allow them to sit in the refrigerator so they will become somewhat firmer. A few minutes before they are to be served I take them out, roll them in some fine cracker crumbs, and fry in deep 375° fat for about 3 minutes. All the vegetables I formerly used are now cut into a tossed salad where they belong, and served with, not in, the Crab Cakes.

CRAB IN WHITE WINE

To 1 cup of dry, white wine add ⅛ teaspoon nutmeg, ⅛ teaspoon cayenne, and ½ teaspoon salt. Place over low heat and when warm

* See the chapters "Sassafras for Food and Drink" and "Wild Rice: Epicurean Delight" in *Stalking the Wild Asparagus*.

add 2 cups crab meat. Simmer slowly 10 minutes, stirring frequently, then add ½ cup light cream. Cook and stir 10 minutes more, then remove ¼ cup of the juice and beat with 2 egg yolks. Add this yolk-and-juice mixture gradually, stirring all the time, and keep stirring and cooking for about 5 minutes more, but don't ever allow it to come quite to a boil. Serve on toast triangles, and accompany it with a glass of the same wine used in the recipe.

A COUPLE OF QUICKIES

I am often the dinner chef at my house, and some days I am just too lazy to struggle with fancy creations; that's when I appreciate simple recipes that put a meal on the table with very little effort. At these times I may merely open a can of marinara sauce or meatless spaghetti sauce, mix in 1 cup of crab meat, heat it all, and serve it over boiled spaghetti. With this I might make a tossed salad with lots of watercress and heat up a loaf of crusty French bread. With a little glass of dandelion wine to accompany the main course, and a demitasse with assorted cheeses to finish it, we get by.

One evening I was preparing to serve an easy dinner of Smithfield ham when a fisherman friend dropped by and gave us 4 freshly caught crabs. Four Blue Crabs, even large ones, as these were, are simply not enough for two hungry people, so I decided to add them to the main dish I had already planned. Smithfield hams are selected hams that have been sugar-cured and smoked in the regular manner, then cured in a hot room until they develop a flavor and tenderness no ordinary ham can equal. I sautéed the ham slices in butter over medium-low heat, then removed it and placed it in a warm oven, while I sautéed the meat from the 4 crabs in the juices and butter left in the pan. I spread the sautéed crab meat over the slices of Smithfield ham and served it with raw-mushroom salad. Delectable!

If you hesitate to go crabbing because you are afraid you might fail at cooking these scuttling, pinching, aggressive crustaceans, forget it. Anyone with a little imagination who is willing to exercise ordinary care and pay a little attention to details can produce superb dishes when he starts with such a delicious ingredient as crab meat. So get yourself some ring nets, folding traps, or just some bait and lines, and go out after those crabs.

5. Razor Clams and Other Marine Cutlery

O NCE I had the opportunity to talk with a biologist whose specialty is Invertebrate Paleontology. He has spent his life studying the fossil remains of the myriads of creatures that inhabited ancient seas and have since disappeared. When I got enthusiastic about some of the marvelous adaptations to environment one sees in nature, he remained unimpressed. "Of course nature has produced some wonderful adaptations," he said, "because she has tried almost every conceivable possibility."

The various genera and species that we ordinary folk lump together under the common names "Razor Clam" and "Jackknife Clam" are good illustrations of this willingness of nature to try almost anything. The Surf Clam and the Pismo Clam have adapted to wave-beaten shores by developing rock-hard shells that can be tightly closed, but the Razor Clams, often found on the same beaches with their heavy-shelled cousins, have thin, fragile shells permanently open at each end. To escape both wave damage and predators, the Razor Clam depends entirely on its remarkable ability to burrow underground out of harm's way. In the same punishing environment, one species has learned to take the blows without damage, while another has learned to dodge them entirely.

Unlike other burrowing creatures, the Razor Clams do not progress through the sand by digging. No dirt is thrown or carried from its burrow; the sand is merely pushed aside and compacted enough to allow the slender body of the Razor Clam to pass. Although ordinarily inhabiting more or less permanent burrows, the Razor Clams have the ability to progress through wet beach sand much as fish pass through water or birds through air, although the mechanism used is vastly different in each case.

The Razor Clam has adapted the part we call its "foot" into an amazingly dexterous and useful member. Most ordinary bivalves have short, wedge-shaped feet, but not the Razor. It can protrude its foot beyond the end of its shell for nearly its own length, and can change the shape of it at will for various functions.

To propel itself downward through the sand, the Razor Clam first thrusts down its probing foot, shaped for this purpose like a sharpened pencil. This sharp-pointed instrument easily penetrates the sand for several inches. Then the point begins to swell, and in less than a second, the end of the foot has become a round disk pressing into the surrounding sand and firmly anchoring the clam. Then the powerful muscles of the foot contract, pulling the clam down to this disk where it wedges itself tightly with its two shells while the foot resumes its pointed shape and is thrust downward again.

If you want to see this process in operation, lay a Razor Clam on a wet beach and watch it. First the foot comes forth and is thrust deeply into the sand. Then, with a sudden jump, the elongated clam leaps upright, balancing over its anchorage, and immediately starts sliding down into the sand, and it can disappear completely in only a few seconds. This is a good subject for your movie camera with a close-up lens, and will furnish pictorial proof that not all clams are the lethargic, inactive creatures they are usually thought to be.

When a Razor Clam wants to return to the surface, it simply reverses the process, first forming an anchoring disk just outside the lower end of its shell, then thrusting its body upward. The disk is then relaxed and the foot drawn up for another step.

By speeding up this latter process, the Razor Clam can also swim through the water at a good rate of speed, although its directional control is a bit erratic, probably because of its curved shell. Once I stepped into shallow water to rinse the sand from four Razor Clams

I had just captured, and absentmindedly laid them down on the bottom, as one could safely do with other kinds of clams. Instantly, each clam formed a disk just outside its shell, then my clams shot off in four directions, each pursuing a zigzag course for some distance before dropping to the bottom and starting to dig in. By the time I had retrieved one of them, the other three had disappeared forever. When swimming, the Razor Clam can uncoil that versatile foot with the speed of a steel spring being released, and each thrust carries the clam three or four feet through the water.

For the forager who gathers his own food by the sea, the various Razor Clams will furnish some of the finest prizes that will ever fall into his hands. Because one never sees fresh Razor Clams in American markets, many people do not know that they are edible. The reason the Razor has not become a market clam in the fresh state has nothing to do with its excellence, but is due to its keeping qualities. The gaping shell can't be closed, so the clam dries out and dies a short while after being removed from the water-bearing sand in which it lives. One of the Pacific Razors, *Siliqua patula,* is canned and sold under the euphemistic label of "Ocean Clams," a name I never heard used for this species in the area where they abound. For those of us who like to catch our own seafood and eat it the same day it is caught, there are few creatures that will furnish more tasty dishes than the Razor Clams.

THE ATLANTIC RAZOR CLAM
Ensis directus

This common Eastern Razor Clam is found in sand in the intertidal zone from Labrador to the west coast of Florida. The average specimen is about 6 to 7 inches long and about 1 inch across, but I have taken a few that measured more than 10 inches in length. The shell is actually white, but in life it is covered with a glossy greenish or olive periostracum, which is the thin, horny covering of the shell proper. This elongated clam is slightly curved lengthwise, and the edges are parallel, the ends gaping and squarish, with rounded corners. No one who has seen it will question how the Razor Clam received its common name. These long, slender bivalves so closely resemble a fine old straight razor with horn handles that one expects to be able to open a blade.

Actually, you don't have to open a blade from the Razor Clam to cut yourself severely on it, as I once had occasion to find out. I was raking Quahogs at low tide when I noticed a number of little rectangular holes near the water's edge, betraying the presence of a colony of Razor Clams. A rake is useless against Razor Clams, so I went after them with my bare hands. Stepping softly to a hole I would quickly thrust my fingers into the sand around it and clamp down. About one time in three I would feel the clam between my fingers, and although I couldn't pull it out, I could hold it while I dug with the other hand until I could dislodge the anchor it had made of its foot. Before the tide came in, I filled my pail with delectable Razors, but in the process I struck the sharp edges of those thin shells so many times that the tips of my fingers were covered with fine cuts that became so sore that I couldn't operate a typewriter; I had to take a week off and go fishing so they could heal. It's a tough life!

A spade is little better than a rake for taking Razor Clams. Sometimes, at extremely low tides, one will see a whole colony of Razor Clams protruding from their holes an inch or two, resembling a number of small stakes driven into the beach. At these ideal times, it may be possible to walk softly to such a group and suddenly shove your spade under one or two clams, cutting off their retreat, and flipping them out on the beach, but even then you are likely to cut more clams in two than you capture.

My bare-handed method proved painful, but I was on the right track. After my fingers healed, I went back to that same beach armed with a garden trowel and a pair of stout kitchen tongs. Approaching each hole quietly, I would make a quick thrust, and most of the time I would have the clam between the jaws of the tongs before it could escape down its burrow. When you try this, don't attempt to tug the clam loose from its efficient, disklike anchor. If you get stubborn about it, so will the clam, and you'll end up pulling the foot right in two, leaving both you and the Razor Clam looking pretty silly, since that foot is the largest piece of edible meat in the clam. Just maintain a gentle pull on the tongs to prevent the clam from going deeper, and use the garden trowel with the other hand to dig the holding ground away from the anchor.

All this commotion will cause the rest of the Razors in the immediate vicinity to scuttle down their burrows, but if you just sit quietly for a minute or two, you will soon see one of those little rectangular holes start overflowing with sandy water, a sure sign that the Razor Clam is coming back up. Grab it with the tongs and repeat the whole process. Because Razor Clams tend to live in thickly populated colonies, one can sometimes fill a pail with these fine eating clams from a few square yards of beach.

A friend in Portland, Maine, wrote me of the weird method he uses to catch Razor Clams. He carries a box of ordinary table salt to the beach, and when he finds a group of the little square holes that tell him of the presence of Razor Clams, he puts about a teaspoonful of salt into each of several holes, and then waits a few minutes. He claims that the salt will invariably cause the Razor Clam to surface. I know this sounds like the old gag about sprinkling salt on a bird's tail to catch it, but it actually works on some beaches. After experimenting with this technique, I have concluded that this saltwater creature doesn't like extra salt in its burrow, but that is no guarantee that salt will cause it to surface. I was puzzled for a while, because this trick seemed to work in some areas and fail in others. I finally came to the conclusion that salt will cause the Razor Clam to surface where there is a hard clay or rock subsoil under the sand, but if the beach is sand all the way down, it merely burrows deeper to escape the unwanted salt. Try it on the beach you visit and see what happens.

Recently I developed another device that promises to be the easiest method of all to extract these elusive clams, and this rig will work on many of the other species mentioned further on in this chapter. This is no more than a ⅜-inch iron rod about 3 feet long with a hook fastened to one end and the other end bent to form a handle. Adjust the length to your own height, so it can be carried like a walking cane. The bottom of the hook is sharpened just enough to make it easy to thrust into the sand. In use, this sharpened bottom is pushed into a burrow until you feel it touch the clam, then it is quickly slid past the clam and his probing foot. The handle is given a quarter twist, bringing the hook under the clam; then, as you pull up, the point of the hook first dislodges the anchoring foot, then catches in the gaping lower end of the Razor's shell, and three times out of four the clam can be brought to the surface with very little effort and no digging whatever. Any blacksmith or welding shop can shape such a hook for you, and the cost will usually be very reasonable.

The best way to prepare Razor Clams for cooking is to wash the shells thoroughly, then cover the clams with clean seawater into which are sprinkled several handfuls of cornmeal. Let them stay in this bath for two hours, and they will void most of the sand and replace it with the cornmeal.

There is no end to the way Razor Clams can be eaten. Use them in Clambakes, Bouillabaisse, Clam Newburg, or in almost any other clam recipe found in this book. To steam Razor Clams, put them into a covered kettle and add only ½ cup of water, and steam until all the clams are open, about 10 minutes after boiling starts. Half a cup of dry, white wine, used instead of the ½ cup water, will give both the clams and the broth a delicious wine flavor. The juice will begin flowing from the clams as soon as they start steaming, furnishing plenty of liquid for the operation.

Steamed Razors need no further preparation to be delicious. Remove the meat from the shells, and dunk each bite in a savory Melted-Butter Sauce (page 21). Serve the broth scalding hot in demitasse cups. Or you can remove the Steamed Clams from their shells and use them and the broth in Clam Pies, Chowder, Stuffed Clams, or Clam Spaghetti Sauce.

To make a delicious Clam Pie, first make a rich crust of 2 cups flour, 1 teaspoon salt, and ⅔ cup shortening. Sift the flour and salt together. Cut the shortening into the flour with a pastry blender until pieces are the size of rice grains. Sprinkle 1 tablespoon cold water on the flour, toss gently with a fork, then push the dampened part to one side of the bowl. Sprinkle another tablespoon of water over the dry part of the flour, and keep repeating until all is dampened, but don't use more than 7 tablespoons of water, or your crust will be too wet to handle properly.

Gather up the dampened flour in your hands and form a ball. Divide in half, and roll each part on a floured surface to ⅛ inch thick, rolling from the center toward the edges. Fit one round of crust into pie plate as bottom crust, and use the other as top crust.

Melt 3 tablespoons of butter over medium heat, chop 1 large onion into the melted butter, and sauté it until it becomes clear and yellow. Stir 2 tablespoons of flour into the butter until it is smooth, then pour in ½ cup clam juice and ½ cup milk, and cook until it thickens. Stir in 2 cups chopped, steamed Razor Clams and a dash of freshly ground black pepper. Pour into bottom crust, fit top crust, perforate in a few places to provide an escape hatch for the steam, and bake in a 400° oven for about 20 minutes, when it should be nicely browned. This will serve four people or two gluttons.

Razor Clams make excellent chowder. Just follow the recipe for Quahog Chowder given on page 21, substituting chopped, steamed Razor Clams for the chopped Quahogs.

The Razor Clam's shell is not the right shape for making Stuffed Clams, but there is no reason why you cannot stuff Razor Clam meat into the shells of Quahogs or Surf Clams, even those empty shells that can be picked up along the beach. Scrub the shells thoroughly, make the stuffing as in the directions on page 22, substituting ground Razor Clams for the Quahog called for in that recipe, and put it into the buttered shells. Serving Razor Clam meat in Quahog or Surf Clam shells may border on nature-faking, but it does no harm to their delicious flavor.

For some purposes, the Razor Clam should not be steamed, but should be removed from its shell raw. To remove the shell from a live Razor, slide a thin knife down the seam opposite the hinge and sever the adductor muscles. Then the clam will relax. Force the shell open, and cut the meat from its moorings on each side. There is nothing to remove or throw away but the shells, for the Razor Clam can be eaten entire, like the oyster. Especially, do not reject the dark-green or blackish mass near the hinge. This is not dirt, as some people suppose, but is the clam's "liver," rich in glycogen, or animal starch, and this is the source of the sweet flavor in fried clams.

The meat of the Razor Clam is fine-grained, white, and of excellent flavor. The small, finger-sized ones are so sweet and tender that they need no cooking whatever to be delicious. Slice the meat into bite sizes and eat with lemon juice or Melted-Butter Sauce.

Young, tender Razor Clams make the best Fried Clams of any Eastern species. After removing the raw clams from the shells, pat them dry between paper towels, then shake in a bag with ½ cup flour, 1 teaspoon monosodium glutamate, and a dash of black pepper. Next, dip each clam in egg well beaten with 2 tablespoons cold water, then roll in fine cracker meal. Fry in hot fat, 375°, for about 5 minutes, and drain on paper towels.

Large Razor Clams are apt to be too tough to eat as raw or Fried

Clams. Being such a Fried Clam addict, I found a way around this. I grind the raw meat of large Razor Clams and fry it as Clam Fritters. Sift together 1 cup flour, 1 teaspoon monosodium glutamate, and a little black pepper. To this add ¼ cup clam juice, caught when you removed the Razors from their shells, ¼ cup milk, and 2 beaten eggs. Stir until well blended, then add 2 cups ground clam meat and 1 tablespoon melted butter. Drop by spoonfuls into 375° fat, fry 4 to 6 minutes, and drain on paper towels. Served with tartar sauce on toasted rolls and with a good cup of coffee, these fritters are a seaside version of heaven.

THE STOUT RAZOR
Tagelus gibbus

Tagelus is usually called the Stout Razor on the Atlantic coast, but along the Pacific where there are several closely related species, they are called "Jackknife Clams," a better descriptive term. The Eastern Stout Razor is common from Cape Cod to Texas, living in dense colonies in stiff mud or muddy sand between the tides. Nearly every seaside stroller is familiar with Stout Razor shells, which litter the beaches in many localities, but so elusive is this burrowing clam, that, despite its abundance, few people have seen the living creature.

This is an easy species to recognize. It is shorter and wider than the Razor, an average specimen being about 4 inches long and 1½ inches across. The sides are nearly parallel and straight, rather than being curved. The shell is heavier than that of the Razor, gaping at the ends and rounded, rather than being squared off. The hinge section is near the middle of one side, and there are 2 small teeth on each valve. Dead shells found on the beach are usually a dull, chalky white, but in life the shell is covered with a glossy, yellowish-brown periostracum.

All clams have two siphons, one to draw in water laden with food and oxygen, and the other to discharge the spent water and metabolic wastes. On many species, these two siphons are united, making what is commonly called the clam's "neck," and one can see that it actually has two channels only by close examination or dissection, but on the Stout Razor the siphons are separate throughout their length, giving this unusual clam 2 necks. These siphons are long and slender, and can be extended about the length of the clam,

 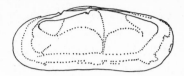

Above, Stout Razor Clam. *Exterior and interior of shell. Size: about 4 inches long, 1½ across. Below,* Ribbed Pod. *Length: about 1½ to 2 inches.*

 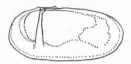

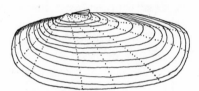 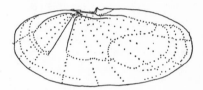

West Coast Razor Clam. *Size: about 4 to 6 inches long, 1½ to 2 inches wide.*

Left, Bay Scallop. *Size: about 2½ to 3 inches across.*

Right, Thick Scallop.

so the Stout Razor usually rests about its own length under the sand and extends its siphons upward to the surface through 2 separate holes about 1½ inches apart, and it is these pairs of twin holes that betray the presence of Stout Razors to those who know how to look for them.

If you examine the siphons of a Stout Razor closely, you will see some tiny orange-colored spots about the openings. These are called *ocelli*, and are really primitive eyes. I doubt that the Stout Razor can see much detail with these eyespots, but it can tell night from day and is wary of sticking its siphons outside the burrow during daylight hours when hungry sea gulls are flying about looking for just such juicy tidbits. There must be individual Stout Razors with poor eyesight, or else with courage that amounts to foolhardiness, for sections of *Tagelus* siphons are sometimes found in the crops of sea gulls.

The reason the Stout Razor so easily eludes a clam digger armed with a spade is that it lives in a permanent burrow, extending 18 inches or more downward. This burrow is lined with a secretion from the clam's mantle section, and by using its versatile foot in the same manner as does the common Eastern Razor, the Stout Razor can slide up and down this slippery hole with great speed. Nor is it simply a matter of being willing to dig deeper, for when you start digging at the top of the burrow, the clam simply disappears in the soft sand or mud at its bottom, and the Stout Razor can travel through this medium much faster than you can dig through it.

Although the Stout Razor is active and clever as a clam can be, it is not invulnerable. With the clam hook described earlier in this chapter, you can not only capture a specimen to examine, but can easily get enough to make a delicious Clam Chowder or Clam Pie. When you see a pair of twin holes that indicate its presence, thrust the hook into the mud between the holes and into the burrow. Slide it down until you strike the clam, then quickly slide the point past the shell and probing foot, turn the hook under to dislodge the foot, catch the barb in the lower end of the gaping shell, and draw the clam to the surface.

The Stout Razor is a perfectly edible clam that is rarely eaten, because of the difficulties of capturing sufficient numbers. The flesh is of good flavor, and while it is not quite as good as young Eastern Razors for eating raw, or frying, it makes excellent Fritters. First soaked in clean seawater and cornmeal, then steamed in the shell, the meat can be eaten as is or used in any of the recipes that call for the meat from steamed clams.

An unusual kind of spaghetti sauce can be made of almost any kind of steamed clam and its broth. Even those who think they dislike spaghetti may discover on eating this white sauce enriched with sweet clams that it was the overcooked tomato paste and parmesan cheese of ordinary sauce of which they were not fond, and not spaghetti itself.

To make this Clam Spaghetti Sauce, steam enough clams to yield 2 cups of clam meat and 2 cups of broth. In a saucepan heat 3 tablespoons olive oil and saute ½ cup minced onion and 3 crushed cloves of garlic. When these turn yellow, smooth in 2 tablespoons flour and brown slightly. Add the clams and broth and stir and cook until slightly thickened. Serve over boiled spaghetti. A 1-pound package of spaghetti and this sauce recipe will furnish generous servings for four people.

THE RIBBED POD
Siliqua costata

From Nova Scotia to North Carolina, one may occasionally find this pretty little shell among the drift on the beach. From 1½ to 2 inches long, it is thin and fragile, oval-elliptic, moderately elongated, and broadly rounded at each end. The color is greenish yellow tinged with purple, and the surface of the shell is smooth, polished, and often iridescent. The hinge is not in the center of the clam's back edge, but is away off down toward the foot. Inside the shell, there is a rib extending from the hinge part across the valve, which helps to strengthen the fragile shell. This is a neat, attractive little shell, and one that is eagerly sought by shell collectors.

Only once did I ever secure enough Ribbed Pods to sample their edibility. A group of us were just "fooling around" on a sand beach, trying to see what strange life-forms we could turn up. The tide was exceptionally low and the sea so calm that we had no trouble launching a car-top boat right from the open beach. I had a wooden box with a ¼-inch mesh hardware-cloth bottom, which I secured just outside the transom of the boat. The sea bottom was shallow for a long way out at this point, so even after getting several hundred yards offshore, we found the water still little over waist-deep. I went overboard and began shoveling sand from the bottom into the box, while my companion poured water into it to wash the sand through the mesh bottom and strain out any shells. By pure luck we struck a bed of Ribbed Pods and soon had several quarts of them.

I had never tasted Ribbed Pods, nor had I ever heard of anyone eating them. However, I did know that two other species of *Siliqua*, living clear across the country on the West Coast, are not only edible, but are considered to be the finest clams in that area, and this is a high rating, for our Pacific shores harbor some very delicious clams.

We steamed a few Ribbed Pods and found them of excellent flavor. Then we opened some raw, rolled them in cracker crumbs, and fried them. Superb! With the remainder we made a pot of chowder that was pronounced delicious by all who tasted it. We concluded that it is only the difficulty of securing these little clams in large enough quantities that prevents their being great favorites among seafood connoisseurs.

THE WEST COAST RAZORS
Siliqua patula and *S. lucida*

Strangely enough, the famous Razor Clam, or Ocean Clam, of open Pacific beaches is more closely related to the little Ribbed Pod described above than it is to the common Eastern Razor. From far up in Alaska to Lower California, this is one of the most highly appreciated clams of the West Coast. From Monterey, California, to Alaska is the home of *Siliqua patula*, and from Monterey southward is found *lucida*, but there are only technical differences between these two closely related species, and to us ordinary folk they look alike, so one description will do for both.

This is a very different clam from the one we call a Razor in the East. It is shorter and broader, on the average, and has rounded ends, instead of being abruptly squared off as is the Eastern Razor. This Western Razor reaches a length of from 4 to 6 inches and a width of from 1½ to 2 inches. The foot is large and powerful, and the siphons are short and united except at the very tip. The shells are thin and fragile, only slightly arched, and the periostracum is glossy and varnishlike in appearance.

Razor Clams are found only on broad, flat, pure-sand beaches that are exposed to the open sea. There are few places in California where this clam is abundant, partly because there are few suitable habitats for it along that steep coast, and partly because the range of the tides is less than it is farther north. Wherever found in California, it is greatly appreciated and jealously guarded by the local population. Strict laws prohibit the sale or shipment of Razors, and may even forbid digging for home or camp use by nonresidents, so check the local game laws before you start digging for Razor Clams.

In Oregon, Washington, British Columbia, and Alaska, this clam really comes into its own. There it is dug commercially and the product is sold as canned or frozen Ocean Clams, a good name, as these clams are never found on the beaches of inland bays or estuaries, but always where the deep-water rollers break over them. In addition to the commercial diggers, the local residents, campers, and tourists get their share of this succulent, sweet clam, and many maintain that one taste of this delicacy will destroy one's taste for all other species of clam. It *is* a good clam, but even after enjoying it many times, I still retain my catholic taste for seafoods of all kinds.

The only way I have ever taken Western Razors was with a spade, as my tong and hook methods have been developed since my last trip to the Pacific Northwest. As this clam does not live in a semi-permanent burrow, but merely slides here and there through the loose, wet sand, I doubt that the clam hook would be much good against it, but I believe the kitchen-tong-and-garden-trowel method would work.

This Razor is never found on high, well-drained beaches, but always on pure-sand beaches that are flat enough to retain a substratum of water, and where there is no mixture of gravel or rocks in the sand that might break that fragile shell or impede its movements. The clam rests vertically just below the surface with its hinge side toward the waves and with the very tip of its siphon protruding into the air. It is so sensitive to any vibration in its neighborhood that it will disappear at the first sound of a step or a spade. When first disturbed, it retreats to only about its own length under the sand, but if it hears pursuit it will go deeper, and it can also evade a digger by traveling horizontally as well as vertically.

When using a spade, one walks softly across the wet beach looking for the projecting tips of the Razor Clam's siphons, which are very hard to see until the eye is trained. Another way is to smooth the sand, where the presence of a Razor Clam is suspected, by tapping it with the flat of the spade. This will cause the Razor Clam to move downward, and as it goes a small pit will appear in the smoothed sand. Either way you have one chance, and only one chance, to capture that particular clam. A quick thrust of the spade will cut off its retreat and throw it out on the surface. If you miss on the first thrust, it is all but useless to continue digging, for your chances of overtaking this speedy underground traveler are very slim indeed.

Once you have the clam out on the surface, it must be immediately secured and put into a container, or it will soon disappear again. When dropped on wet sand, the powerful foot goes into its digging routine with movements so swift the eye can hardly follow them. A Razor Clam has been observed to flip itself upright and completely disappear below the surface in seven seconds.

The Western Razor Clam is so famous as Fried Clams that some people never think of preparing it any other way. However, this precious delicacy can also be eaten raw or prepared according to almost any clam recipe you can find or devise, and still be good. I am looking forward to using it in a real, old-fashioned, New-England-style Clambake on my next visit to that grand country, the Pacific Northwest. I hope to fill the steaming trench of Rockweed with Western Razors, sweet Washington Butter Clams, Dungeness Crabs, boneless strips of king salmon each wrapped with a sliver of onion in a corn husk, and maybe a mighty Geoduck as the *pièce de résistance*, surrounded by new white potatoes and tender young sweet corn. I'm hungry already.

THE JACKKNIFE CLAMS
Tagelus californianus

Nearly everything that was said about the appearance and habits of the Eastern Stout Razor, *Tagelus gibbus*, could be repeated about this closely related Western species. It is found from Santa Barbara south into Lower California, and is especially abundant around San Diego Bay. Like the Stout Razor, it lives in colonies between the tides, in permanent lined burrows built in mud or mixed mud and sand. It is not used to any extent, for the same reason that the Stout Razor is not eaten—because these clams are smarter than most clam diggers, and can seldom be collected in sufficient numbers to make cooking them worthwhile.

The Jackknife has fine-grained, white, sweet flesh, and readily yields to the clam hook, so I suspect that its future will not be as secure and peaceful as its past has been. I have a report that small boys capture them now with a local version of the clam hook and sell the clams to fishermen to use as bait. The slender foot of a Jackknife does make good fish bait, but unless the clams were in plentiful supply I doubt that I would allow a fish to eat such a tender, sweet tidbit. I once bought a few live ones from a bait shop, took them to our cottage, and cooked them. While not quite as delicate and delicious as the Eastern Razor, the flesh was of good flavor and would make excellent fritters or chowder.

Colonies of Jackknife Clams can be located in the same way as those of the Eastern Stout Razor: by observing the many pairs of twin holes where they thrust their separated siphons to the surface. Make yourself a clam hook, and enjoy some good sport and fine eating on your next visit to Southern California.

ANOTHER JACKKNIFE
Solen sicarius

Superficially, this clam looks much like *Tagelus* and is called by the same common name, but the two differ greatly in details. *Solen sicarius* is smaller, on the average, than the clam described above, and has a very glossy periostracum that gives the shell a yellow tinge. The siphons are united instead of separate, and the hinge is located not in the center of one edge as on *Tagelus*, but away down toward the foot end, a feature that identifies this genus at once. The range of these two clams overlaps in the south, but that of *Solen sicarius* extends northward to British Columbia. It is seldom found in abundance, but where it can be located, it can be taken with the clam hook and used in the same ways as its kinsmen.

From Santa Barbara south, over the same range as *Tagelus californianus*, one occasionally encounters a little brother to the above clam, this one called *Solen rosaceus*, and commonly named the Pink Jackknife, although its small size and delicate build would make "Penknife" more appropriate. This is a little fellow only about 2 to 2½ inches long, with a rosy-hued shell; otherwise it is much like *sicarius*. Often found living in colonies of *Tagelus*, this little clam, when taken with a clam hook, can be mixed with the larger game and eaten with no discrimination.

To sum up: the various Razor and Jackknife Clams are widely distributed, not impossible to capture, and well worth the effort. They represent a large resource of fun and delicious food that has hitherto been mostly neglected by the seaside vacationist and camper. Here is a group of clams smart and agile enough to make outwitting them a real challenge, and delicious enough to make eating them a rare taste thrill. You're really living when you search out a hidden clam colony by the water's edge at extremely low tide, then stalk and outwit this elusive prey, and top off your day with a steaming plate of chowder or fritters, knowing that you, alone, procured the main ingredient.

6. Blue-Eyed Scallops— and Others

(*Pecten* species)

NO MATTER how you think of them, scallops are superlative. The rounded, fan-shaped shells with their radiating ribs, waved edges, and 2 winglike projections at the hinge line seem to be the most esthetically pleasing of all sea shapes. Artists through the ages have incorporated the scallop shell in a thousand designs. Once, a soldier-pilgrim picked up a pretty shell from the shores of the Mediterranean and soon the scallop shell was recognized as the pilgrims' official badge. One who wore it was hailed as a Crusader who had been to the Holy Land. In this way it entered heraldry, and scallop shells appear on the devices of many famous families. In modern times, a great oil company has adopted this pleasing shape for its symbol and has dotted the roadsides of the world with signs in the shape of a scallop shell.

The biologist points out that the scallop is the one of the most active of all bivalves. The gourmet acclaims it as one of the most delicious of seafoods. This lowly shellfish has even added to our language. A pleasingly waved margin is called a "scalloped edge" and a recipe baked in an oven dish with bread crumbs is called a "scalloped" dish. The latter received the name because the ancient housewife did not throw away the pretty scallop shells after removing the tasty mollusks they contained, but kept them to use as plates, bowls, cups, or baking dishes. I am glad to see modern housewives rejecting the ersatz plates of artificial materials and returning to the original scallop shells as a handy and artistic way to serve baked seafood dishes. Of all the scalloped dishes ever prepared, none have surpassed what was probably the first of them all, Scalloped Scallops.

Scallops are worldwide in distribution, and there are over two hundred species, but they all have enough family resemblance to be easily recognized. The common Eastern Bay Scallop, *Aequipecten irradians*, found from New England to Cape Hatteras, is a fairly typical species. A bit on the small side, when compared to some of the huge deep-sea scallops, the Bay Scallop is from 2½ to 3 inches across as a full-grown adult, but this is large enough for it to be of commercial importance, and thousands of pounds are dredged up each year and sold on the market. It is almost circular in outline, with 17 to 20 ribs and 2 equal "ears" forming a straight hinge line about two-thirds the width of the shell. It habitually rests on one side, and the upper valve is more convex and darker than the lower one.

The scallop never crawls or burrows, so its foot is tiny and slender and apparently is used as a manipulative member rather than as a means of getting around. It has no siphon tubes whatever, as it never goes underground where it has to reach up for food and oxygen. The scallop swims through the water with good speed and some directional control by rapidly opening and closing its valves and squirting water through openings in its mantle cavity. The scallop had perfected jet propulsion for its own use ages before man even appeared on this globe.

The scallop's mobility enables it to escape danger, to seek out water that is rich in oxygen and foodstuffs, and to distribute its species widely. No one who has watched scallops flitting about the tide pools and shallows will deny that they have an instinct for play as well as for survival.

They seem to express the joy of life also by the colors they wear. I cannot tell you what will be the color of the first Bay Scallop you will capture. On my desk are the shells from twelve scallops that I took alive from a cove on Buzzard's Bay. No two of them are alike in color. They range from white to black, but on the way they detour through all the browns, tans, reds, purples, and steely blues you could name. Several have zones, or concentric bands, of lights and darks, and others have rays of several colors spreading out from the hinge line to the outer margin.

The fringed mantle, usually visible between the partly opened valves of live specimens, generally matches the shell in color, and those with rays of different colors will show sharply demarcated areas of different color on the mantle. Around the edge of the mantle there is a row of as many as 50 bright, shining eyes, dots of iridescent green encircled by rings of turquoise blue. Biologists tell us that these are real eyes, having cornea, lens, choroid coat, and optic nerve, and they more closely approximate vertebrate eyes than any found among bivalve mollusks. One would think that with this many eyes, this mobile creature could easily see where it is going, but since the scallop swims by jetting water through openings in the mantle it necessarily moves backward, hinge first, so the eyes are looking in the opposite direction.

Instead of having two adductor muscles to close its valves as do most other bivalves, the scallop has only one, but this one is proportionately huge. It is this muscle that is the scallop of commerce, and few people need to be told how eminently edible it is. The scallop doesn't "clam up" when captured, but beats its valves wildly in an attempt to escape, sometimes giving the finger of its captor a severe pinch. It is far from being a symbol of silence, as is the oyster or the clam; I have sometimes held a newly caught scallop in each hand and laughed at the comic symbol of two old gossips exchanging the latest dirt amidst a great clacking of false teeth.

This disinclination to keep its mouth shut has made the scallop unsuitable for marketing in the shell, for it soon loses all contained water and dies. It has become the practice to remove the muscle at the fishing grounds and throw the rest of the creature away. This is a criminal waste of good food, for, like the oyster and the clam, the entire scallop is edible, even delicious, when freshly caught and properly prepared.

Because of its mobility, the scallop is likely to be found on a variety of bottoms around the world. Ordinarily it prefers sheltered shores. The vicinity of good scallop beds will always be betrayed, however, by the abundance of dead scallop shells along the neighboring shores. When these are plentiful, the only problem is locating the live ones. The best places to hunt are in patches of eelgrass, and I know several places in New England, New York, and New Jersey where it is pos-

sible to pick up enough scallops for a good meal just by wading out
in the shallow water at low tide. At other places it is better to drift
slowly over likely spots in a rowboat or canoe, searching the bottom
with a glass-bottomed box and scooping up with a long-handled net
any scallops seen. Probably the most productive way of all to go after
scallops is with a snorkel mask and a pair of swim fins. This is very
elementary skin diving, as one merely floats along over shallow bottoms with one's face under water and scoops up with a net any
scallops that are found. Carry a bag tied to your swimming trunks
to receive your catch.

Many coastal states limit the size and number of scallops that can
be taken, but these provisions are usually generous. There are also
likely to be laws regulating open seasons, and some states require a
license, so check the game laws of the state where you intend to go
scalloping.

Scallops are the easiest of all bivalves to prepare for the pan. Just
slide a sharp knife between its valves and sever the edible muscle
where it joins one valve. It will then open wide and can easily be
cut from its moorings. You can remove the adductor muscle and eat
it separately, or you can cook everything you find in the shell together.

The secret of superb scallop dishes is to allow the shortest possible
time to elapse between the time this delicious shellfish is taken from
the water and the time it appears on the table. I remember one
autumn day when I waded ashore with a net bag of scallops tied to
my bathing trunks to find unexpected guests at the beach cottage
we were renting. One of the guests had brought a present of a three-
pound box of beautifully white cultivated mushrooms. I would have
preferred wild mushrooms, but only to protect my reputation as a
nature cook, for I must admit that these cultivated *Agaricus* are hard
to excel. Within minutes of the time that I emerged from the water,
I was opening the still lively scallops, removing each entire creature
and dropping each one into an inverted mushroom cap. Over each
scallop I placed a small slice of bacon, and very soon a great panful
was ready to be slid under the preheated broiler. When the bacon
was crisply done, these Broiled Scallops were brought to the table,
the sizzling still audible. Less than an hour had elapsed since these
delectable shellfish had been taken from the water, and my guests
swore they had never tasted better scallops.

Everyone is familiar with Fried Scallops—the muscles, that is—but the entire scallop can also be fried and will rival Fried Oysters any day. Cut each one loose from the shell, leaving the hinge muscle still attached to the body. Dip in egg that has been beaten with a little water, then roll in fine cracker crumbs. Fry in deep fat heated to 375°, drain on paper towels, and serve hot with a wedge of lemon.

Just to prove that the entire scallop is delicious, I sometimes remove the round muscles, serve them separately, and make Scalloped Scallops of the rest of the meat the shell contains. For this purpose a pint of scallop meat, either with or without the adductor muscles, is chopped coarsely and mixed with ½ cup cream, salt and pepper to taste, 1 cup cracker crumbs, and a small dash of nutmeg. Scallop shells are filled with this mixture and covered with white bread crumbs that have been stirred with all the melted butter they will take up. Bake at 350° for about 25 minutes, or until the buttered bread crumbs are nicely browned, and serve immediately.

Some seafood fanciers think that fresh scallop muscles are better when not cooked at all, and I am inclined to agree. When removing freshly caught scallops from their shells, I like to have a little dish of melted butter near so I can sneak a few of those sweet scallop muscles when no one is looking. Sometimes, when I am out foraging the seashore at low tide, I become so interested in the pursuit of scallops that I forget to go in for my lunch, but I don't go hungry. I just open a few scallops on the spot and eat their sweet muscles with no sauce but their own juicy, sea-given saltiness. This enables me to cheat a bit on the game laws, for I never count the scallops that are already safe inside me in my daily bag limit.

When serving Raw Scallop Muscles to your guests, chill them for an hour in the refrigerator, then serve them on a bed of lettuce or watercress ringed with small, red, cocktail tomatoes cut in half. With them served Melted-Butter Sauce (page 21) or seafood cocktail sauce. Or you can spear the muscles on toothpicks and serve them as hors d'oeuvres or cocktail snacks.

For those who prefer their scallops cooked a bit, there are few dishes better than Scallop Muscles Sautéed in White Wine. Melt 4 tablespoons butter in a saucepan with a tight cover, then add 1 pint of the muscles and 1 tablespoon finely chopped fresh parsley. Sauté until the scallops are golden, add ½ cup dry white wine, and serve on hot buttered toast.

South of Cape Hatteras, one finds the very similar Zigzag Scallop,

Pecten ziczac. These are small, from 1½ to 2 inches in diameter as adults, with 18 to 23 ribs on the shell, and, like the Bays, their "ears" are practically equal in size. Zigzags are even more colorful than the Bays; they sport many pinks, soft blues, yellows, and other hues, and in design they affect zigzag lines and mottling as well as concentric bands and colored rays. This scallop is probably too small ever to be commercially important, but that is no excuse to neglect gathering them for our own use. Even in areas where they are abundant and easily secured, many local people do not seem to know that this delicious little shellfish is edible. I have seen areas in the South where the numbers of these little animated rainbows seemed unlimited; there, one could easily have compensated for their small size by gathering more of them. Besides, this scallop isn't so small when one knows that the entire animal is edible, and not just the adductor muscles.

In southern Florida and around the Keys, one finds other, tropical species of scallop, and around the Gulf Coast one encounters still others. The only thing to remember about scallops is that *all* species are edible, and while size and quality vary slightly from species to species, none of them would be called poor food by anyone whose taste buds are in good working order.

It was out on the West Coast that I first discovered the possibility of gathering my own scallops, and I have retained a soft spot in my heart for those tasty Pacific species. Early in the Great Depression, I left New Mexico and went to California with the first wave of the now famous "Okie" migration. Being little more than an overgrown boy, I insinuated myself into a family of dust-bowl refugees from western Kansas. We followed the fruit and vegetable harvests with varying fortunes and camped out under the stars. One evening we stopped by a tidal sand flat that was part of a system of shallow bays and sloughs south of Long Beach. As the tide was then receding, I waded out to see what this shore had to offer that was edible.

I realize now that this shore must have been teeming with edible life, but at that time my eyes had not been trained to see it. I was walking along the shore, still empty-handed, when I heard a splash such as fish sometimes make when trying to avoid being trapped in a tide pool. I soon found the shallow tide pool and saw that all this commotion was being made by two big scallops trying to flap back into deeper water to keep from being stranded by the receding tide.

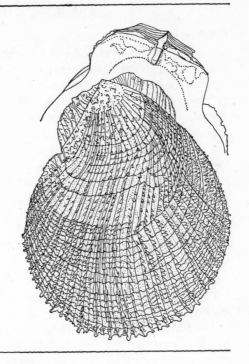

Above, Northern Scallop. *Exterior and detail of interior of shell. Size: about 2 inches in diameter.*

Right, Rock Scallop. *Exterior and detail of interior of shell. Size: about 4 to 8 inches.*

Steamer Clam. *Size: about 2 to 3 inches long; occasionally, 4 to 5 inches.*

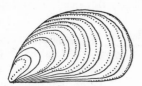

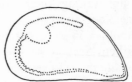

Blue Mussel. *Size: about 3 inches long.*

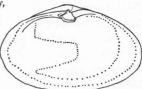

I quickly secured these two, and thus alerted, I began to find others, some lying on the sand that was already bare, and more on the sand that was still covered by a foot or so of water. In a short while I had filled a half-bushel fruit hamper with scallops. I have never since gotten so many scallops so easily. Call it beginner's luck.

I bore my treasure back to the camp in triumph. I still remember that our camp stove was a regular cast-iron kitchen wood stove. It was a ridiculously heavy piece of equipment to haul about in our old cars, but we would set it up with a length of stovepipe beside the road, or at the end of an orchard, and turn out delicious meals with some very unpromising materials. Tonight, however, the materials were anything but unpromising. I shucked out the scallops while my temporary "mother" built a hot fire in that incongruous stove, and soon an appetizing aroma filled the still evening air as the scallops were fried to a turn. We ate this heaven-sent manna from tin plates, and I thought that I had never tasted better food. And so it was that I discovered how superior scallops are when cooked as soon as they are removed from the water.

The scallop that so providentially furnished us with a good dinner was the Thick Scallop, *Aequipecten circularis*. It is thick, as scallops go, and the ones that fed us that night were from 2½ to 3 inches in diameter. While not quite as unrestrained as its Eastern cousins in the matter of color, *circularis* is not exactly drab. The colors run more to muted browns, tans, and yellows, but many of them have attractive zigzag patterns near the outer lips. They are found from Monterey, California, to Peru, but only in shallow, sheltered waters where the tidal currents are not too swift. I'm sure the tidal flat where I first met this delicious bivalve has long since been filled to make room for some new industry or housing development, but I'll wager that this fine scallop still finds sanctuary in the many small bays, sloughs, and lagoons along the Pacific coast of Southern California and Mexican Lower California, and that many good scallop banks are just waiting for the knowledgeable forager or seaside camper to come along and claim his share of them.

Another West Coast scallop that is dear to my memory is the Northern Scallop, *Chlamys hindsii*, that I have often gathered on rocky shores of Puget Sound. It is about the size of *circularis*, but has finer and more numerous striations on its shell, which is usually pink in color. It lies on its right side, and the left or upper valve is always darker-colored than the lower one. This scallop has a tremendous north and south distribution, being found from the Bering Sea to San Diego. It is a profitable market scallop and commercial trawlers are constantly sweeping the bottom for it wherever they can go. For this reason we always tried to find, for our amateur scalloping, flats so studded with rocks that these commercial scallopers steered clear of them. I have picked up a goodly number of these scallops from tide pools and rocky shallows by just wading out at extremely low tides, but this is not a very dependable way of taking them. Using a snorkel mask and swim fins in protected, rocky bays is probably the most productive way of gathering these scallops for home use, but one has to be part polar bear to enjoy the cold water of the Northwest Coast. In calm bays one can drift about in a rowboat among rocks, where the larger scallop trawlers would never dare to go, searching the bottom through a glass-bottom box and scooping up scallops with a long-handled net. This may not be the most efficient method, but one can stay dry and comfortable while getting a mess of scallops. Perhaps some of the new scuba diving gear with its warm dress would be the best thing to use, but it does seem a bit elaborate just to get enough scallops for dinner.

The scallop, like the Mussel, has a gland, called the "byssal," from which it can extrude anchor lines at will. When a young scallop gets tired of swimming about or being carried by the tides, it spins some threads, winds them into a cord with its clever little foot, and attaches itself to a seaweed or rock. However, unlike the Mussel, it doesn't settle down there for the rest of its life, but casts off and abandons its anchor line as soon as it is seized with a new desire for freedom, and resumes its merry way. Most species, as they grow older, gradually lose this byssal habit and elect the free life permanently. There is one West Coast species, however, that seems to deplore the roving, carefree existence of most of its kinsmen. This scallop has worked out a way of life so at variance with that of other scallops that biologists put it in another genus and call it *Hinnites giganteus*. Commonly it is called the Rock Scallop or Rock Oyster, but since the latter common name is shared by several genera and many species of unrelated seashore creatures, none of which are Oysters, I fear that such a name will be of little help in identifying it.

The Rock Scallop starts life very much like any *pecten*, but soon tires of this endless flitting about and settles down. It selects the half-buried under-surface of some rock and cements one of its valves firmly to it, and there remains for the rest of its life. It continues to grow and may become as much as 6 inches in diameter, making it the largest scallop that you are likely to find in shallow water, but it rapidly loses its native beauty after giving up its freedom. The lower shell grows to fit its rocky foundation, and the upper shell grows to fit the lower one. It becomes hardly recognizable as a scallop, but all its life it carries, up near the hinge, a plain imprint of the pretty shape of its shell before it started to grow awry. It also retains some of the habits and instincts of the free-swimming scallops, and when approached it will often flap its valves as if it had forgotten that it was permanently anchored. Inside, the shell is white, with a large purple stain up near the hinge. The mantle is dark-colored, but the large adductor muscle is creamy white and perfectly delicious.

This scallop is found from the Aleutian Islands to the southern end of Lower California. I have collected a good mess of Rock Scallops in Monterey Bay, but it reaches its highest development and is most abundant along the bay-indented and island-protected western coast of British Columbia. I'll never forget a boating-camping trip a group of us took up the famous Inside Passage. Many of the un-spoiled bays we explored are inaccessible by land, and we camped on completely unspoiled beaches. After the new moon, there were about four days of exceptionally low tides baring seldom-seen lands that were veritable seafood markets except for the prices. We were all seafood fanciers, and for a week we hardly touched the stores we had brought along. When the tide was out, the table was set. We feasted on Native Oysters, Butter Clams, Horse Clams, and Geoducks. We reveled in Blue Mussels, Moon Shells, and Boring Clams. We gorged ourselves on Blennies, crabs, and Bent-Nose Clams, but most of all we relished the Northern Scallops and Rock Scallops that could be collected in abundance at any low tide. The Rock Scallops were clustered so thickly on the rocks just below the low-water mark that we could get all we wanted by just knocking them loose with a stone held in the hand, and never had to get wet above the knees.

One woman in our party was an artist with herbs and taught me some new and wonderful ways of cooking shellfish. Once she sautéed the huge muscles of Rock Scallops for only a few minutes in butter flavored with crushed celery seeds. Heavenly! Another meal was made of minutes-fresh raw scallops served with a savory dill sauce. But best of all were the Rock Scallop muscles that were threaded on skewers, brushed with cooking oil, sprinkled with crushed anise seed, and broiled only about 4 minutes over charcoal. Unbelievably good! With such variety in seasonings we didn't tire of this steady diet of seafood. We even ate fried Blennies and scallops for breakfast. I'm sure the half-dozen people who shared that wonderful expedition and tasted all that outstanding food will forever remember it as one of the most pleasant vacation trips of their lives.

7. The Common Periwinkle

(*Littorina littorea*)

THE Edible Periwinkle is seldom eaten in North America, where it crowds rocky shores from the Arctic to New Jersey. In England this delicious little gastropod has been better received. In Charles Dickens' day "winkle shops," little restaurants that specialized in serving Periwinkles, were common about London's famous East End. Periwinkles, roasted in the shell, were hawked about the streets much as roast chestnuts are sold in New York and Philadelphia. A 100-year-old record states that at that time it required 1900 long tons of Periwinkles per year to supply the London market, and these were valued at 15,000 English pounds. At the rate of exchange then prevailing, this would come to about two cents per pound, so Periwinkles were not the exclusive right of rich gourmets, but could be enjoyed by everyone. The street hawkers and winkle shops may be gone, but the Periwinkle is still one of the most common edible mollusks sold in England, and one can still buy them by the pound, measure, or bushel in Billingsgate Market.

Why did the Britons who came to this country stop eating Periwinkles when they arrived here? The answer to that one happens to be easy. There were no Edible Periwinkles on our shores when the early settlers came. The only Periwinkles the Pilgrim Fathers found on the "stern and rockbound coast" of New England were several species too tiny to be used as food. Then a strange thing happened. About 1850 the exact species so highly esteemed in the British Isles appeared on the coasts of Nova Scotia and New Brunswick. Whether it had been purposely or accidentally introduced, or whether it had worked its own way around the Arctic, seems not to be known. By 1868 it was reported from the coast of Maine, and by the end of the century it had reached Cape Cod. The Cape was considered for some time to be the southern limit of these delightful little gastropods, but they soon jumped that barrier and appeared along the coast of Long Island. Now they are found as far south as the mouth of Delaware Bay, and they may be still moving.

Meanwhile, another species, the *Littorina irrorata*, commonly called the Lined Periwinkle, the Marsh Periwinkle, or the Southern Periwinkle, and formerly known only from the Gulf of Mexico and Florida, began moving north. Today these Southern invaders have reached as far as the coast of Massachusetts, so already the two species overlap. Our coasts, once devoid of Edible Periwinkles, are now amply supplied, but we, poor souls, have, during the barren interval, forgotten how to make use of them.

The Edible Periwinkle reaches about 1 inch in diameter and approximately the same in height when mature. In shape the shell is a squat cone with 7 or 8 whorls. The commonest color is olive, but there is great variation, some being gray or brownish yellow, others banded with dark red or reddish brown, and some appearing entirely black.

The Southern Periwinkle resembles its Northern relative in size and shape, but in color it is unmistakable, being a yellowish white with rows of tiny brown spots arranged in spiral fashion up its whorls. Both species are good to eat, but I prefer the Northern Common Periwinkle both for flavor and because it is more easily removed from its spiral shell. The Edible Periwinkle is usually found on rocky coasts exposed to the surf, but the Southern species prefers salt marshes somewhat protected from wave action.

Another reason I prefer Edible Periwinkles is that they are usually easier to gather in quantity. There are any number of places along the East Coast where Periwinkles by the bushel could be gathered if one had any use for so many. On rocky, weedy coasts they can be seen crawling by the dozens over the seaweeds when the tide is low. In such places it is easy to gather enough for a meal in a very short while.

While forgetting how to eat Periwinkles, we have forgotten, of course, how to prepare them for the table. The Periwinkle is no convenient bivalve that merely needs to have a muscle severed to render it helpless; it is a corkscrew-shaped creature in a spiral shell closed at the opening by a close-fitting limy door called an *operculum*. To remove it from this twisty shell, boil it a few minutes in water to which a handful or two of salt has been added. This salt is not for seasoning, but serves to shrink and toughen the meat of the periwinkle until it can be "unscrewed" from the shell in one piece. Don't overcook them; as soon as the operculum drops off and the animal protrudes from the shell far enough to be grasped, it is done. As soon as it is cool enough to handle, twist each morsel from its spiral prison with a pin or nutpick. It is not a difficult operation, and compared to picking out crab meat, it is not even tedious.

Periwinkles can be eaten with no further preparation; I have dined well on nothing but boiled Periwinkles with crusty French bread and butter and a glass of dry, white wine. These same tidbits, speared with a toothpick and dipped in cocktail sauce or melted butter, make grand hors d'oeuvres or cocktail snacks.

As a main dish, or as one of the features of a mixed seafood platter, Periwinkle Fritters are excellent. To 2 cups of cleaned Periwinkle meats add ½ cup flour, 1 teaspoon monosodium glutamate, ½ teaspoon baking powder, and 1 egg well beaten. Mix until the flour is evenly dampened, then wet the hands and form the stiff batter into little cakes. Fry these in a well-greased frying pan until golden brown on both sides, and serve with a salad made of wild seaside plants.

A little fancier main dish is my version of *Omelette à l'Arlésienne*. Beat together 8 eggs, 1 six-ounce can tomato paste, and ½ cup water. Season with 1 teaspoon celery salt and a pinch of marjoram. Stir in 3 tablespoons chopped chives, and crush a clove of garlic over the mixing bowl, adding only that part of the garlic that passes through the fine holes of the crusher. Now, lightly brown 1 cup of cleaned Periwinkle meats in 2 tablespoons butter. Pour in the egg mixture, and cook the omelet over low heat. The secret of a good omelet lies fully as much in the cooking as in the ingredients. An even slightly scorched omelet isn't fit to eat; with proper heat it can be cooked so the top is just set when the bottom is lightly browned. At this point, make a crease across the omelet with a spatula and neatly fold it over, then roll it directly onto a hot platter without trying to lift it from the pan. A narrow trail of paprika will brighten the top side. Garnish the dish with 4 small sprigs of parsley. This dream of an omelet will serve four people luxuriously.

8. The Surf Clam:

ABUNDANT AND DELICIOUS

(Spisula [Mactra] solidissima)

THE Surf Clam, or Hen Clam, is the largest bivalve found along the Atlantic seaboard, and one of the most common, being found in great abundance at the level of the lowest tides on most sandy, surf-washed beaches from Labrador to Cape Hatteras. It grows to a length of 7 inches, and specimens 6 inches long and 4½ inches broad are not at all uncommon. The outside of the roughly triangular shell is ridged with concentric striations, or growth rings, and in life it is covered with a light-brown periostracum, but this has usually been rubbed off those dead shells that litter sandy beaches at the high-tide mark. These are the shells seashore visitors most often carry home to use as ashtrays, and I have a large one serving that purpose on my desk right now. You can be sure it is a Surf Clam if you see a spoon-shaped depression on the inside just below the apex of the triangular shell. This was the attachment of the internal ligaments that held the shell together when the creature was alive. Even in the dead shells, one can also see the smooth, circular scars where once were attached the muscles that enabled the clam to open and close its valves. These muscles are the finest morsels produced by any bivalve, and they are the prizes I seek when I go surf clamming.

Sure, I know that old clam digger told you Surf Clams are only good for fish bait, and unfit for human food. The books on the subject are no more reliable, one saying, "occasionally used for food, but usually regarded as too tough or too sandy;" and another erring on the other side by declaring, "It is not as tender as the hard clam, but nevertheless is eaten and is especially sought after for clambakes."

I have tried including Surf Clams in Clambakes, and find the sandy viscera and the muscular foot, which is tough as a piece of harness leather, very poor fare. If the Surf Clam is merely substituted for the proper species in chowder and Clambakes, you will not like it, but any seafood connoisseur knows that each species requires individual treatment to be at its best. The Surf Clam can furnish some of the finest seafood ever eaten, but only if it is approached with the right spirit, reliable information, and the proper skills and techniques.

As in the case of the scallop, it is only the *adductor muscles* of the Surf Clam that delight the gourmet. These little cylinders of tender muscle, only about an inch in diameter and the same in height, are smaller than those of the scallop, but are even better in taste and texture, and the Surf Clam has two such muscles while the scallop has only one.

In some places Surf Clams can be gathered at any low tide, but they are gathered more easily if you time your clam-digging excursion for a day or so after the full or new moon, when the tide has the greatest run in and run out. At the ebb of these spring tides, sand spits, bars, and peninsulas appear that have not been above water for a fortnight, and it is beneath such seldom-seen bits of land off sandy beaches that you will find the Surf Clam most plentiful. Sea gulls know the best clamming spots, so if you see them settling in great numbers on some spit or sand island as soon as it appears above the water, go out and give the birds some competition.

The Surf Clam has a very short siphon that is a combination breathing tube, food inlet, and waste-disposal unit. Because of the shortness of this muscular tube, the Surf Clam doesn't burrow deeply as do some other species of clam, but is usually found just under the surface of the sand. If the clams are widely scattered, or if you have to delve for them with a fork or spade, you are in the wrong spot. In a good Surf Clam bed, a hand thrust into the sand at random will almost certainly encounter a large clam just under the surface. From the hole made as one clam is lifted out, it is easy to move the hand through the yielding sand until another clam is found. This is not digging clams or hunting clams, it is merely gathering clams as one harvests the fruit from a bountiful tree. On good surf-clamming ground most of the clams brought up will be 5 or 6 inches long, and anything less than 4 inches long is not worth the bother, as they are too tedious to clean and the yield of edible meat is too small to make them worthwhile. Leave the smaller clams in their sandy bed, and this will guarantee there will be large ones there next year, and the next.

I usually open Surf Clams right at the beach. This is partly to avoid the ridicule that would result if the family saw me clean a bushel of clams and get only a quart of edible meat, and partly so the discarded parts can be fed to the sea gulls, avoiding a garbage-disposal problem. I owe those birds something for disturbing them at their feeding grounds, and the show they put on while being fed is definitely part of the pleasure of a surf-clamming expedition. As soon as the gulls discover that free food is being distributed, they hover above us in flocks, exactly balancing on the breeze and filling our ears with their raucous begging. A discarded piece of Surf Clam thrown among them will be unerringly caught in mid-air by the nearest gull, who will then wheel away from the rest of the thieving crowd until he has gobbled it down. This aerial exhibition of grace and gluttony is always enjoyed by any guests I have along, and children, especially, find these performing gulls a very delightful part of a day at the beach.

The right tool for opening clams is a thin knife with a rounded end to the blade. On the posterior end of the clam, where the siphon protrudes, there is a narrow opening between the two valves into which a clam knife can be inserted. Push the knife forward, being careful to keep the blade tight against the bottom shell, and slide it under the near adductor muscle, severing it where it fastens to the shell. This eases the tight grip with which the clam keeps his shell closed, so you can slip the knife around the opening between the two shells and slide it under the other muscle. Now the clam relaxes completely and is easily opened. I pull the valves apart over a funnel thrust into the neck of a jug, to catch the copious, bluish clam juice. Two quick strokes with the clam knife sever the adductor muscles from the other valve, and the two meaty morsels can be lifted out with the fingers, for they are lightly attached to the rest of the clam. These little "Beach Scallops" are far more tasty than any Cherrystone Clam on the half shell I ever tasted, and are delicious eaten raw right on the spot.

You can also enjoy the liquid part of your haul right at the beach. Build a fire of driftwood, put a camp kettle on the coals, and dump a quart of clam juice into it. As soon as it boils, add a quart of milk, which you thoughtfully brought along. As it reaches a simmer, pour the steaming brew into mugs and pass it around. I call this simple recipe Clam "Chocter" (combining the words "nectar" and "chowder"), and I not only use it as a warming beach refreshment, but often serve it indoors before a seafood meal. Always serve it in cups rather than bowls, for it is drunk, not eaten with a soup spoon. You can add a sprinkling of freshly ground black pepper to smarten the flavor, but on the beach I enjoy it without added seasoning.

Only two bite-sized pieces of meat and a little clam juice from each of those huge clams! Wasteful? No more so than scallops, of which it takes a half-bushel to yield a quart of the adductor muscles we eat, and what the Surf Clam lacks in quantity it makes up for in quality. A great many more of these clams die each year from storms, shifting sands, and the depredations of the voracious gulls than are taken by the few clam diggers who know their worth. When the Surf Clam reaches about 7 inches in length, it seems to die simply because the purpose for which it lived has been fulfilled. It is nature that is wasteful, and we are only harvesting our share of her overflowing bounty.

Too tedious? Cleaning Surf Clams is not nearly as tedious as picking the meat from boiled crabs, and it is fully as worthwhile. I find that I can gather and clean enough Surf Clams to yield a quart or two of muscles during one low tide, and have a gallon or so of clam juice as a bonus. Where else can you be certain of obtaining that much delicious seafood in so short a time?

Some say the best way to cook these sweet muscles is . . . don't. These are the folks who prefer them raw, without even the addition of sauces or seasoning to mask their delicate flavor. But not all of us are such purists. I like them both raw and cooked, with or without sauces. Never wash these muscles in water. Shake them in a jar with some clam juice decanted from the top of the jug; this will remove any clinging sand without diluting the natural clam flavor as water would. Drain the "Beach Scallops" well and chill them thoroughly, then eat with cocktail sauce, Melted-Butter Sauce (page 21), or just eat. They make the finest of teatime snacks, hors d'oeuvres, or seafood cocktails.

For Fried "Beach Scallops" wash 1 pint of Surf Clam muscles in clam juice, drain well, then rub dry with a clean cloth. Shake the muscles in a paper bag with ½ cup flour and 1 teaspoon monosodium glutamate until they are evenly coated. Fry gently in 4 tablespoons butter, being careful not to get the pan too hot. About 10 minutes' cooking should turn them into golden bites of the best seafood you ever tasted.

Ordinary Clam Chowder is plebeian, but when it is made from the juice and adductor muscles of Surf Clams it is truly epicurean. Cut 4 slices of bacon small, and dice 1 large onion. Fry these together until lightly browned, then add 2 cups of diced potatoes, 1 dried leaf from a Bayberry bush, cover all with about 3 cups of clam juice, and boil until the potatoes are done. Add 3 cups of rich milk, ¼ teaspoon freshly ground black pepper, and bring just to a simmer, not a boil. Make a thickening by blending 1 tablespoon flour and 1 teaspoon monosodium glutamate with a little milk, and add this to the chowder a little at a time, stirring until the soup thickens slightly. Add 2 cups of clam muscles that have been washed in clam juice. Heat through, but do not boil; serve hot with saltines and a good tossed salad that includes watercress. If, by a miracle, some is left over, it can be reheated next day and will be even better.

Hot Clam Nectar is merely clam juice heated to the boiling point and served in small glasses or demitasse cups. Some find it a bit too salty, but remember, this is natural sea-salt, loaded with many trace minerals your body needs for glowing health.

An excellent Clam Juice Cocktail is made by combining 2 cups raw-clam juice, 2 cups tomato juice, 2 teaspoons Worcestershire sauce, and the juice of half a lemon. Chill well before serving.

If you find a good bed of Surf Clams and get so carried away that you gather more than you can immediately consume, they will keep perfectly well in the freezer and can be used when it is too cold or inconvenient to go to the beach. Just put the washed clam muscles in a container, cover with clam juice, and freeze. Because of the high salt content, frozen Surf Clam muscles will keep better than most frozen seafood, and months later these frozen morsels will be almost indistinguishable from the freshly dug product.

Because so few people know their worth, there is little competition for Surf Clams, and crowded beds of these succulent, bivalves can often be found near heavily populated areas where the supplies of more popular clams have been sadly depleted through over-digging. While you are gathering them, someone is sure to come by and tell you that Surf Clams are not edible. Don't get into an argument about it. Just be pleasant and keep your secret. Leaves all the more for you and me.

PUBLISHER'S NOTE:

In his review of *Stalking the Blue-Eyed Scallop* in the *New York Times* on June 1, 1989, long-time outdoor writer Nelson Bryant contributed the following on the Atlantic surf clam:

"Gibbons wrote that only the sweet-fleshed adductor muscles of the clams were suitable for human consumption and described how, when opening the bivalves on the beach, he fed the remaining parts to waiting herring gulls. That is a sin.

"One can set the adductor muscles aside if one wishes, either to consume them raw or to use them in recipes suitable for bay or sea scallops, but they and the other parts of the clam—siphon, stomach, stomach contents, mantle and huge muscular foot—make a superb base for a pasta clam sauce after being run through a meat grinder. Ground-up surf clams can also be used for clam fritters."

9. Transcendental Seafood Combinations

A TRULY artistic cook is not one who makes a massive collection of cookbooks and recipes and then slavishly follows them to the letter. A good cook is creative, and the success or failure of his creations depends on knowledge, skill, judgment, experience, and to some degree, on talent. Even this last mystical gift can be analyzed to some extent. A talent for cooking must be composed partly of a good memory for tastes. Some people possess the ability to recall a flavor so vividly that they can mentally taste a food just by thinking of it, and they can project a number of these mental tastes into original compositions and predict with a high degree of accuracy how the new dish will turn out. If the imagined dish seems to lack something when it is mentally combined, these culinary artists can reach into their taste-memory banks and come up with exactly the right herb, flavoring, or ingredient to bring it to perfection.

Craig Claiborne, the famous food editor of *The New York Times*, calls such people "born cooks," implying that this talent is a natal one. But is such a gift entirely a grace, capriciously bestowed on few and withheld from many, or is it to some degree a learned skill? Aldous Huxley, in his book, *Island*, tells of a people whose grace before meals consists of focusing the attention, with total awareness, on the first bite of food taken. Perhaps we have here a hint on how to go about acquiring this marvelous gift.

If you would like to try to develop this skill in yourself, and at the same time test your aptitude for this gastronomic art, try preparing some simple food, say a single Cherrystone Clam broiled on the half shell. Place it on a table, seat yourself before it, and relax, dismissing all the thousand and one other things that clamor for your attention, and focus your whole being on that one little clam. Look at it. Imagine how a number of them could be combined with other foods and garnishes to make a dish attractive to the eye. Savor its aroma. Would the addition of another food fragrance improve that appetizing smell? Put the single succulent bite in your mouth, and pinpoint your awareness on what you are doing. Sense its temperature, and decide whether it would be better warmer or cooler. Feel its texture and the amount of resistance it gives to your jaws as you chew, and judge whether these are entirely pleasant experiences or otherwise. Taste it with every taste bud you possess, detecting each element of the composite flavor. Finally, feel it as it slides down your throat; then sit awhile in silence and contemplate the whole experience, setting every detail firmly into your mind.

Hours later, when the odor and taste of the clam have completely disappeared, repeat the whole experience with another food. This time try a single bite of bacon, broiled to a turn. Go through the entire process. Concentrate on that bite of bacon; hear it sizzle; see its pattern of fat and lean; breathe its hickory-smoked fragrance, and taste its salty savoriness to the utmost.

Next day, preferably when it has been several hours since you have eaten and you are beginning to feel hungry, find time to sit quietly alone and recall those two experiences. Sink into the memory of eating that clam, and as nearly as possible, actually relive the episode. See it, smell it, feel it in your mouth, taste it, yes, and even hear its final hiss as it was taken from the broiler. Do the same for the bacon. Now comes the final test. Recall both tastes with the utmost possible vividness, then combine them. Would these two flavors, so different from one another, go well together? Wonderful! We have just reinvented Clams Casino, a noble dish, for which you will find a recipe on page 21.

Another element of this so-called talent of the "born cook" is love. I know that sounds like a banality, but I am convinced that most really fine cooking is motivated by love. We have all heard of the bride who worshiped her husband and set burnt offerings before him three times a day, but if this happened in real life I would suspect there was considerable unconscious hostility in that worship. A cook who does not love the task, the utensils, the ingredients, and the finished product is not likely to advance far in this demanding art. But especially must we love those members of our families, the friends and guests, and even "the stranger within our gates" who will consume the food we so painstakingly prepare. While it is probably true that a good cook would, from sheer force of habit, produce an edible meal, even if he were cooking for his worst enemy, it is only love that can motivate us to give the infinite attention to detail that produces the truly superb dish.

ART, SCIENCE, AND BOUILLABAISSE

One of the functions performed by great artists, often unintentionally and unknowingly, is to give new and wonderful meanings to formerly commonplace words. When this happens, these words can no longer be translated into other tongues, and must be borrowed intact by those who speak or write in other languages. Generations of French chefs have performed this service for the word *bouillabaisse*. To translate this literally into English and have it come out "bass soup" or "fish stew" borders on blasphemy.

Thousands of touring Americans have tasted this dish in its native land, and dozens have wheedled recipes for this gastronomic creation from famous French chefs. Many of these have found their ways into newspapers, magazines, and cookbooks over here. The most remarkable thing about all these recipes is the way they vary so widely from one another. Often two recipes are so different that one wonders if the authors are writing about the same food. Many of them are too complicated to follow, or call for unobtainable ingredients. I have one before me that begins, "Take one blackfish, one sea bass, one small striped bass, a medium-sized live lobster, an eel, two soft-shell crabs . . ." I'm discouraged already. How is one ever to assemble exactly these ingredients, besides the many more exotic components mentioned further on in the same recipe? Another cooking expert, apparently in an effort to make this dish conform to the habits and cooking abilities of the average American housewife, calls for all frozen or canned ingredients—and even includes canned tuna. Ugh!

The truth is that the chefs who furnished the originals from which these recipes were adapted would laugh loud and long at the very idea that the creation of a Bouillabaisse could be reduced to a mere matter of exact recipe and precise directions. To these artists a Bouillabaisse is a matter of inspiration, dedication, and genius. The list of ingredients that a French chef gave to some traveling food editor wasn't a recipe, but was merely a description of the Bouillabaisse he was currently serving, and he probably never dreamed that this list would be touted about as *the* authentic recipe.

He knows that the Bouillabaisse of the French Riviera is quite a different dish from that served in English Channel ports. Beyond that, the Bouillabaisse of one chef will not be the same thing as that of another, and the same chef will make quite a different Bouillabaisse in the spring from the one he so proudly served in the fall. The seasons in which certain ingredients are obtainable come and go, but an ingenious chef can create a delicious Bouillabaisse any time of the year. Furthermore, one of these true artists, brought to this side of the Atlantic, could, if given time to become acquainted with our excellent seafoods, compose a Bouillabaisse the equal of any ever served in France.

Essentially a Bouillabaisse is several kinds of seafood, often as many as twelve varieties, cooked with vegetables, wine, herbs and seasonings, and served over toasted French bread. Having said this, I am reminded of the set of directions that came with an oil-painting kit designed to be used by amateurs. These said, "Set up the canvas on the easel. Squeeze some paint from the tubes onto the palette, then apply the paint to the canvas with a brush. This is the method used by Rembrandt, Whistler, Van Gogh, and all the other great masters."

This analogy can be carried further. I think it fair to say that a Bouillabaisse constructed according to an exact recipe and detailed directions is related to a truly inspired dish in precisely the same way that a painting made by filling numbered spaces with similarly numbered paints is related to an inspired oil by a talented artist.

After such a diatribe, I am in no position to give a specific recipe for Bouillabaisse. However, the beginner needs at least a base from which to improvise, so a list of general ingredients and a set of hints on procedure become necessary. To make a Bouillabaisse for six people, I begin with a basic recipe that almost overuses the figure 3. I am no numerologist, and there is no particular magic in this number. An examination of my recipe will reveal that all those 3's were only obtained by juggling the units of measurement until 3 produced the right quantity. It's a mnemonic device that helps prevent mistakes, and it makes it very easy to alter the recipe when you must serve more or fewer people. One only needs to substitute the figure 2 when there are four diners, 4 when there are eight, and so on.

BOUILLABAISSE BY THE RULE OF THREE

3 pounds white-meated fish
3 cups crustacean shellfish meat
3 dozen molluscan shellfish
3 leeks, white parts only
3 cloves garlic, crushed
3 tomatoes
3 teaspoons dried sweet peppers
3 tablespoons olive oil
3 tablespoons melted butter
3 gills dry, white wine
3 cups fish stock
3 sprigs parsley
3 sprigs thyme, or ½ teaspoon dried thyme
3 sprigs basil, or ½ teaspoon dried basil
3 peppercorns, or ½ teaspoon ground black pepper
3 dried Bayberry leaves, or 1 bay leaf
3 small pinches saffron
You will also need a long loaf of crusty French bread,
but I'll be darned if I'll say, 3 one-third loaves of bread

At our house, a Bouillabaisse is always a sequel to a fishing or shore-foraging trip, and the ingredients must change to conform to my catch. Striped Bass, Sea Bass, Red Snapper, and Flounder are all fine fishes to use, but there are many more fine food fishes along our shores that could grace a Bouillabaisse with dignity and worth. It is better to use two or three kinds. I always fillet the fish as described on page 228, and use the heads and bones to make the Fish Stock (page 283) called for in the above recipe.

For crustaceans use crab, shrimp, or lobster, and again, the dish will be better if you use all three. Boil according to the recipes given for each species, and pick out the meat. Dice the lobster and shrimp.

The most commonly used mollusks are Cherrystones, small Soft

Clams, Mussels, and Oysters, but you are certainly not limited to these. Use two or three kinds if available. If you want an authentic French product leave the shells on, and thoroughly scrub the shellfish before adding it to the dish. I usually remove the shells before serving the Bouillabaisse unless I'm trying to impress someone who has eaten this dish in France.

If you can't get leeks, use the white parts of 6 or 8 scallions or chop up 3 medium-sized onions.

A perfect Bouillabaisse is fully as much a matter of skillful building as it is of the ingredients used. Heat the olive oil in the bottom of a large kettle; then add the chopped leeks or onions, the tomatoes finely diced, and only the parts of the garlic cloves that will pass through the fine holes of the crusher. At the same time, melt the butter in a frying pan and gently sauté the fish fillets, which should be cut into 2- to 3-inch squares. This must be done over low heat, or the butter will scorch and ruin the dish.

If you have timed things just right, the fish will be done just as the leeks or onions start to turn yellow. Gently lift the pieces of fish into the kettle with the leeks and tomatoes, and immediately add the 3 cups of Fish Stock. If you use fresh herbs tie the thyme, basil, and Bayberry leaves in cheesecloth and drop them into the soup, to be removed just before serving. Put powdered herbs and dried sweet peppers directly into the soup. Chop the parsley and saffron fine and sprinkle over the soup after it is dished up.

Let the fish simmer in the stock for five minutes; then add the diced lobster, the shrimp, and the crab meat. As soon as the pot boils again, add the wine. After the wine is in, never let it really boil again. As soon as it comes to a simmer, drop in the shellfish that are to remain in their shells, and five minutes later drop in those that have been removed from the shells. Three minutes more, and the soup is ready to serve. When it comes to salting, be careful. Shellfish vary in brininess as do the waters from which they are taken. The only safe rule is to depend on your taste, rather than on a measuring spoon.

Slice the French bread and fry the slices to a very light brown in butter. Stand these slices on edge around a huge soup tureen and pour the hot soup inside. Your guests should already be seated and poised for action when this is done. This is very important. The separate ingredients of a Bouillabaisse require different cooking times to come to perfection, and the whole thing can be wrecked if you have to wait ten minutes before dishing it out on the plates. A beautiful tureen is a smart way to present this dish to those assembled, but it must be a very temporary home for your Bouillabaisse if you are to prevent the oysters, clams, and mussels from becoming overcooked in the hot soup.

Unlike most recipes, this set of directions is intended only as a general guide, and is given to be violated. This is only a basic instrument, on which you can play strange and wild tunes. I have eaten a glorious Bouillabaisse in Seattle in which the seafood ingredients consisted solely of halibut cheeks, crab, Butter Clams, and tiny Olympia Oysters. In Southern California I made one with Bean Clams, Pismo Clams, and the little local Swimming Crabs as main ingredients. In Hawaii I used Spiny Lobster, Samoan Crab, a kind of edible limpet called *opihi*, and two of their most delicious fish, the *moi* and the *papio*. I also added 2 cups of sliced Octopus tentacles that had previously been boiled until they were very tender. When my mainland guests became curious about those little disks of delicious seafood, I told them it was from a local mollusk that was highly appreciated by epicures in many parts of the world and gave them its Hawaiian name, *puloa*. This was no more than the plain truth, for the Octopus is a mollusk, related to clams and oysters, and it is highly appreciated by discriminating diners in Polynesia, Japan, France, and many other places in the world where people put fine food before unreasoning prejudices. I only neglected to say that the English name for the food they were eating with such relish was "octopus."

Once in Indiana, hundreds of miles from the sea, I constructed a Bouillabaisse of large-mouth bass, bluegills, freshwater clams, and the cleaned tails of freshwater crayfish, all taken from the same local stream. I won't say it was the best I ever made, but it was much better than no Bouillabaisse at all.

Probably the strangest Bouillabaisse I ever compounded was on the coast of New Jersey. Out of sheer perversity, I decided to make an epicurean dish entirely of the species that most local fishermen and clam diggers threw away. For fish I used Eel and Sea Squab, which is the drumstick-shaped piece of fine fish from the back of a Blowfish or Swellfish. The shellfish I used were Lady Crab, Mantis Shrimp, Periwinkles, Razor Clams, and the adductor muscles from

Surf Clams. This turned out to be a superior dish, and I especially recommend the Sea Squab, Periwinkles, and Surf Clam muscles to Bouillabaisse artists. Eel is controversial. There are chefs who say that Bouillabaisse without Eel doesn't deserve the name, and it definitely does add something to this dish. However, I part company with the French in the way I add Eel to a Bouillabaisse. Sections of whole Eel, sautéed with the fish and served with the bones still in them, are tasty enough, but they are very unhandy and messy to eat. I skin and fillet the Eel (see page 236), then cut the fillets into bite-size pieces and treat them exactly as I do the rest of the fish that goes into the soup.

Not all my tampering with Bouillabaisse is so radical. Once when the amount I had made looked scanty to my hungry eyes, I added 1 cup of finely shaved smoked salmon and the rest of the wine from the bottle. Since then a French cook has assured me that adding smoked fish to a Bouillabaisse is pure heresy, but my guests thought it exceptionally fine.

Of course, Bouillabaisse is not intended to be served as a soup course at the beginning of a meal. It is a whole meal in itself and a very fine dinner it makes. With it I usually serve nothing but a tossed salad made of watercress combined with dandelion crowns or endive to give an appetite-provoking hint of bitter. Use a chive dressing to which the vinegar has been added with a miserly hand. For dessert, try a tray of assorted French pastries with the coffee.

In all these lists of components, I have failed to mention the most important ingredients of all; these are the heart and soul of the chef. One who would create the perfect Bouillabaisse must bring to the kitchen integrity and devotion, and a touch of genius will hurt nothing. However, do not let these rare qualifications scare you away, for another essential ingredient of this heavenly dish is boldness. Temerity can ruin an otherwise good Bouillabaisse, but even worse, it can prevent you from trying one, and that would be the greatest tragedy of all.

HOW TO HAVE A CLAMBAKE

A Clambake is not quite the artistic endeavor of a Bouillabaisse, but like that famous dish it is seldom twice the same thing. The contents of Clambakes vary widely according to the foods available and the skill and preferences of the cook. While it is not as demanding of perfection as the Bouillabaisse, I don't want to give the impression that it is impossible to produce a poor Clambake. I have tried to eat Clambakes where the potatoes were only half done, the sweet corn overripe or kept so long after picking that it was tasteless, and the clams all full of sand. There are no foods so good that they cannot be ruined by poor cooking.

The only essentials for a Clambake are clams, potatoes, and sweet corn. All else is embellishment, but it is just these embellishments that can transform an ordinary Clambake into a gala occasion. Don't try to make a Clambake merely edible; strive to make it a feast that will never be forgotten.

The best way to indicate what can be done with a Clambake is to describe one that actually occurred, I won't say exactly where, for I'd like to find that place uncrowded with human beings and still crowded with clams the next time I return. It was early autumn, the day of the full of the moon, and a minus tide was due shortly after midday. We were three families, fourteen people in all, a goodly number for a Clambake, for this seaside feast is particularly suited for feeding groups a bit too large to crowd into the ordinary dining room. We had chosen our spot because one of our party knew of a mud flat where Soft Clams were plentiful and there was a nearby rock-and-sand beach where we could have our bake.

Some of the inexperienced members of our group had little faith in our ability to feed such a hungry mob from nature alone, and to allay their fears we very unethically stopped by a fish market on the way. I intended to buy a dozen small chicken Lobsters to eke out the clams we expected to find, but inside, in a saltwater tank all by itself, was the biggest Lobster I had ever seen, weighing fully 15 pounds. The children with me immediately gathered around this tank to admire the monster, and the proprietor moved in with a calculating look in his eye. He obviously was worried about whether this overgrown crustacean would become a white elephant on his hands, so he offered it to me at a bargain I couldn't afford to refuse. On an impulse, I bought this one huge Lobster and nothing else, to the great dismay of the adults in our group, but to the great delight of the youngsters.

We made another stop at a farm market and settled an argument about whether sweet potatoes or newly dug white potatoes were better in a Clambake by buying a bag of each. We also acquired a bag of onions and several dozen ears of sweet corn, newly plucked from a late field in the rear of the store.

The clams were there in the promised abundance. As is usual on expeditions of this kind, when the extra-low tide began uncovering new areas, the sea and beach began yielding unexpected treasures. Someone discovered that a mixed mud-and-sand beach near the mouth of the bay was crowded with Cherrystone Clams and Razor shells. Another found a bed of Surf Clams on the outer beach. Some of the more adventurous waded and swam among the eelgrass, catching dozens of the small but delicious Bay Scallops.

In a sandy spot well above high-tide level, we dug a trench 8 feet long, 2 feet wide, and 2 feet deep, and lined it with stones. All hands were drafted to gather driftwood, and we soon had a huge bonfire laid in the trench ready for lighting when the time came. The older children gathered bushels of wet, live, Rockweed (*Fucus*) and piled it on an old canvas tarpaulin where it would be ready for use. Rockweed, or *Fucus*, is one of the most widely distributed of all seaweeds, abounding on all cool, rocky shores. There are many species of this alga, but they all look much alike and all can be used in this kind of underground cooking. When seen at low tide Rockweed is a limp, hanging mass of brown, leathery, ribbonlike plants repeatedly forked. Along the mid-veins and at the tips of the forked stems there are air bladders that enable this plant to float upright when the tide is in. While gathering the weed, they found the rocks clustered with Blue Mussels and crawling with fat Periwinkles, so they gathered a supply of each to add to our already bountiful food hoard.

These community jobs done, the others scattered to pursue various projects and pleasures while two of us worked at getting the food ready for the steaming-pit. We scrubbed all the shellfish and covered the Soft Clams, Cherrystones, and Razor Clams with clean seawater into which we sprinkled handfuls of cornmeal. This is a trick to get these clams to give up their sand and replace it with something more edible. We opened the Surf Clams and Bay Scallops, removed the muscles, and washed them in their own copious juice. The tough feet of the Surf Clams were passed on to the fishermen in our party to use as bait. The potatoes were washed, and bad spots and eyes removed; then the outer husks of the sweet corn were taken off and the ends of the ears examined to remove any lurking earworms. The great Lobster was laid on its back on the picnic table and killed by being split from stem to stern along the centerline. When we forced the tremendous shell open, we found that what we had been calling a "he" was really a "she," for there were great gobs of *coral*, the reddish roe of the female Lobster. This coral was carefully saved to be added to the stuffing. We removed the intestine, the gills, and the sac under the head, leaving a huge cavity to be filled with the stuffing I began to devise. We had too few scallops and clam muscles to provide another dish which would go around in that crowd, so to keep down arguments about who deserved these delicacies, I added them to the stuffing. These with the coral, a few chopped onions, and half a loaf of crumbled bread made enough to fill even that enormous cavity. We fitted the two sides together and tied them with a stout string, and our outsized Lobster was ready to be consigned to the steaming-pit.

The other people in the party went fishing to while away the hours until the Clambake was ready. Two had gone to the outer beach to try their luck at surf casting. A car-top boat we had brought along took three, all it would safely carry, up a nearby tidal stream to try for flounder and perch. Some of the children sloshed across the salt marsh to fish from the banks of the tidal creek, and others were hopefully lowering ring nets and folding traps for crabs from a nearby bridge.

The luck of the fishermen was nothing phenomenal. The tidal creek yielded half a dozen flounders and a few more perch. The youngsters caught a dozen good crabs, and one of the surf fishermen landed a 22-inch Striped Bass, the only bite either of them had all afternoon.

The fish were filleted and the boneless slabs cut into 4-inch pieces. The presence of the corn husks we had removed from the sweet corn gave someone an idea, so we took each small section of fish and wrapped it in a corn husk with a little salt, pepper, and a sliver of onion. The ends of the husks were tucked under and all of the husk-wrapped packages of fish were stacked into berry boxes so they wouldn't come unrolled while steaming.

We delayed lighting the fire until after midafternoon, for one secret of a reputation as an excellent outdoor cook can be the simple device of delaying the meal until appetites are so ravenous that any decently prepared dish will taste like ambrosia. The fire took nearly two hours to burn down to a bed of incandescent coals. The few pieces of wood that were still smoking were removed and the white-hot embers covered with a 4-inch bed of wet Rockweed. Over the Rockweed was spread a length of new cheesecloth. On this we placed the huge Lobster, and around it we scattered clams, crabs, Mussels, Periwinkles, sweet corn, the berry boxes of husk-wrapped fish, potatoes, both white and sweet, and even included the rest of the onions. Another piece of white cheesecloth was spread over all and then heaped high with the rest of the Rockweed. Then we placed a heavy tarpaulin over the mound to seal in the steam and weighted down its edges with shovelfuls of sand and some rocks.

Cooking a Clambake between blankets of cheesecloth was a new wrinkle to me. This was done because while we were planning this Clambake one man confided that his wife was a bit dubious about eating food that had been cooked on nothing but Rockweed gathered from the beach. A package containing five yards of cheesecloth, enough for both bed and cover, was purchased for seventy-five cents. While it probably added nothing to either the cleanliness or the taste of the food, it did make it look better and was therefore more acceptable to squeamish members of the party. It also made the food easier to uncover and kept such small items as Mussels and Periwinkles from getting lost in the Rockweed. All in all, a good invention, and I'm grateful to the hesitant woman who forced us to think of it.

The pit was closed for one hour. After we removed the tarpaulin, willing hands helped remove the top layer of hot Rockweed, and we turned back the cheesecloth blanket, revealing our Lobster, now a fiery red, in a bed of steamed sweet corn, potatoes, clams, and all the other seafood. We cut the strings that bound the Lobster together and let the two halves lie shellside down on the hot Rockweed where it would stay warm. Picnic plates were soon piled high with Lobster, stuffing, and all the other good things from the steaming-pit.

There were containers of melted butter on the picnic table to go with the sweet corn, seafood, and potatoes. The Cherrystones were eaten whole from their now-opened shells, but most of us preferred to discard the tough necks of the Soft Clams and eat only the soft, delicious bodies with melted butter. Unbelievable quantities of food disappeared in the next hour. Only gradually did we unroll the cheesecloth cover, so that seconds and even thirds were still appetizingly hot. We had coffee or beer for the grownups and a bucket of lemonade for the youngsters. I sided with the children for I think lemonade, if not too sweet, an excellent drink to go with a seafood meal.

None of us had ever before seen Mussels and Periwinkles included in a Clambake, but the children had found and gathered them, and it would have been cruel to refuse to include their contribution. Surprisingly, they proved great favorites. The fish cooked in corn husks proved so good that we'll always use the outside husks of sweet corn to wrap the fish on future Clambakes. Even whole buttered onions are unusual items to include in a bake, but their presence in the pit imparted a delightful hint of savoriness to all the rest of the food. It is innovations such as these that can elevate a Clambake into an exciting new experience, if we can get past the idea of following a recipe or a set of directions.

CLAMSHELLS STUFFED WITH MIXED SEAFOOD

One day a friend and I went fishing, promising our wives that we would have a cookout in our backyard that evening with the catch. Most fishermen know such promises should never be made, and, of course, things went wrong. By early afternoon we had caught exactly one little Flounder between us. During our Flounder fishing we had managed to boat four good-sized Blue Crabs, and by going overboard in the mud near a channel island I had found half a dozen Quahogs by feeling about with my bare feet. We finally admitted defeat at the fishing and hurried to the outer beach to try for Surf Clams before the tide came in too far. By the time we arrived, the tide had already turned, and a heavy surf was coming in. We did get four huge Surf Clams by locating them with our feet and then ducking under for them.

On arriving home with this meager catch, we decided that since we hadn't enough of anything to be cooked alone, we would combine our entire catch in a single dish. We boiled the Crabs and picked out the meat; cleaned our undersized Flounder and dropped it into boiling water for only about 3 minutes, then picked the meat from the bones; opened the Quahogs and ground the tough meat in a food chopper, and added the adductor muscles of the Surf Clams. Some slices of bacon were fried until they were crisp and then crumbled into the bowl where the seafood was accumulating. To this we added ½ cup each of onions, green peppers, and celery, all finely minced, and crumbled in some slices of day-old bread to extend the dish a bit and give it the proper consistency. All this was dampened with clam juice, mixed thoroughly, and stuffed into the clam shells, which were then wrapped in two thicknesses of aluminum foil. When these had been roasted for 30 minutes over the charcoal grill, we ate them with forks directly from the shells, and all agreed that we had stumbled onto a seafood dish that could hardly be improved.

SEAFOOD NEWBURG ON CORN CRÊPES

Essentially a Newburg is seafood sautéed in butter, laced with sherry, and covered with a cream-and-egg-yolk sauce, and most people consider it very fine fare indeed. Nearly everyone is familiar with lobster Newburg; most know it is sometimes made of Shrimp, and a few know of the fine Newburg that can be made of crab meat, but how many have heard of making it with Periwinkles or Surf Clam muscles, diced scallops, or even smoked cod or kippered salmon? I can testify that any of these seafoods can form the base for a delicious Newburg and that it will be even better if compounded from several of them. Not all these Newburgs will taste the same—far from it; each is a superbly different dish, but all are delicious, and with only minor alterations in procedure all can be made by the same recipe and served in the same manner.

To serve four people generously you will need two cups of any of the seafoods mentioned, or the same amount of a combination of two or more of any of them. Besides the seafood, you will need ¼ cup cooking sherry, 6 egg yolks well beaten, ½ pint light cream, and some salt, pepper, and nutmeg.

To prepare the seafood: Slice cooked Lobster meat across the grain. Use Shrimps raw, peeled and de-veined, and if they are large, dice them, but if you use small Shrimps, leave them whole. Crab meat is used just as it is picked from the shells of boiled or steamed crabs. Surf Clam adductor muscles are used raw after a thorough washing in clam juice to remove any clinging sand. Small scallops are used raw and whole, but large ones are diced, not too fine. Periwinkles are boiled ten minutes in salty water, then picked from the shells and used whole. Smoked, kippered, or even fresh fish are diced, although you must take care not to let any bones get into the Newburg.

For tasty combinations I like ½ cup smoked or kippered fish with 1½ cups scallops. Crab, Shrimp, and Lobster combine well with one another in almost any proportion that strikes your fancy. Periwinkles or Surf Clam muscles make unusual and delicious Newburgs, but each should be used alone so that other tastes will not mask their smooth, subtle flavors.

To the horror of chafing-dish fans, I always make a Newburg in a double boiler. Melt ¼ cup butter in the top of a double boiler, then place over direct heat and sauté the seafood, being careful not to let the pan get hot enough to scorch the butter. Shrimp, scallops, or fish should be sautéed 10 minutes, Lobster or Surf Clam muscles only about 5 minutes, while crab meat and Periwinkles should barely be heated through. As soon as the seafood has sautéed for the prescribed time, add the ¼ cup of cooking sherry, stir well, then cook 1 minute more. Place the pan on the bottom half of the double boiler, which should be boiling on another burner. Add the ½ pint of light cream and 6 well-beaten egg yolks and stir—and keep on stirring. Don't let anyone tell you that cooking this dish in a double boiler eliminates the need for stirring. That may be true of some foods, but if a Newburg isn't stirred constantly the egg yolks will cook up lumpy, and a lumpy Newburg is unforgivable. Stir until the sauce thickens slightly, then salt to taste and add a little freshly ground black pepper and a tiny pinch of nutmeg.

Most Newburg is served over toast points, making a very good, but rather unimaginative, dish. After once partaking of a superb crab-and-shrimp Newburg at the table of an outstanding cook, and then later hearing a fellow guest describe it as "some sort of creamed fish on toast," I decided that what this dish needed was a more dramatic presentation so that even insensitive and undiscriminating diners would realize that they were being treated to something pretty special.

One idea I had was to roll the solid parts of the Newburg in crêpes, those thin, delicious, French pancakes, but I found that I did not particularly care for this dish in plain flour pancakes. This started a search for the perfect base for Newburg that ended when I discovered what I call corn crêpes, which bring together the wonderfully compatible flavors of fresh sweet corn and seafood. You will need 2 cups of "corn cream," which can be made of from 4 to 6 average ears of freshly plucked sweet corn. Husk the corn, pick off the silks, then split each row of kernels with a sharp knife. Now, with the back of the knife rub down over the ear, forcing the juice and pulp from the inside of each grain. To 2 cups of this "cream" add 2 well-beaten eggs and ½ cup flour sifted with 1 teaspoon sugar, 1 teaspoon baking powder, and ½ teaspoon salt. Mix well, then stir in milk, a little at a time, until the batter is a little thinner than heavy cream.

Use a round, flat griddle with a handle, and heat it until a drop of water will dance around a few seconds before disappearing. Rub the griddle with a piece of bacon to grease it, then drop the batter on with a large serving spoon. Pick up the griddle by the handle and tilt it first this way, then that, to spread the batter thinly and evenly. The cake will cook quickly, so turn it as soon as bubbles appear on the surface and lightly brown the other side. Slide each cake onto an ovenproof dish and with a perforated spoon dip up a spoonful of the solider parts of the Newburg and place on the cake; then quickly roll it up while still hot. Arrange the rolled-up cakes, seamside down in the dish and keep warm until all are finished. Then pour the sauce over the cakes, leaving a dry trail down the center of the row. In this dry spot on each rolled-up cake, make a slight depression with the back of a spoon, then pour 1 teaspoon cognac in each dimple. Light the cognac and bear the dish to the table in a blaze of glory.

HOW TO FLOUT A RECIPE

After writing the first draft of this chapter, I had an occasion to alter drastically my own directions for making a Newburg, or maybe the dish I came up with couldn't even be called a Newburg. We had a houseguest for the weekend, and spent Saturday at a nearby lake. While my wife and our guest swam and sunbathed, I fished. It was early June, that in-between time for freshwater anglers. Bass season had not opened, and the tame hatchery trout had long since been caught by the put-and-take fishermen, a clan to which I do not belong. However, I am not the kind of fisherman who demands trophy-sized fish at every cast, so I went spinning for Bluegills and in two hours caught my limit of twenty-five, all of them between 6 and 7 inches long.

As the day was still young, I took a hike over the hills that surrounded the lake to see what other gifts nature had to offer in this region. On a grassy slope I found ripe wild strawberries, not growing as thickly as I could have desired, but I did manage to pick more than a pint of this fragrant fruit, besides the toll I took between plant and pail, for I simply cannot resist eating wild strawberries as I pick them. While following a stream back to the lake, I came on a lush bed of wild watercress and picked a bagful of the tenderest sprigs.

As we drove home I was extolling the virtues of the sweet-meated little fillets I would get from my Bluegills and asked our guest how she would prefer to have them cooked. She said, "Ordinarily I am not very fond of fish in any fashion, but I was wild about that seafood Newburg you served the last time I was here. Can you make something like that of Bluegills?"

"Why not?" I answered nonchalantly. I had never heard of making Newburg of Bluegills or any other freshwater fish, but I am an inveterate show-off when it comes to cooking. Besides, I would have promised almost anything to please so pretty a guest.

We stopped to buy sweet corn for the crêpes, and when I saw a pile of coconuts on the produce counter it gave me an idea. In Hawaii I had enjoyed some wonderful Polynesian fish dishes flavored with coconut cream. Why not replace the ordinary cream used in Newburg with the creamy fluid that can be squeezed from grated coconut meat? It was worth a trial.

I skinned and filleted the fish and found that nine of them were turgid females, each containing two swollen sacs of yellow roe, large as egg yolks. Egg yolks? I had another brain storm. I had made caviar of Bluegill roe, and I loved the whole roes gently sautéed in butter and served on toast. Why wouldn't this bright-colored roe serve in a Newburg as well as ordinary egg yolk?

The roes were passed through a sieve to remove the membranes, then beaten in a knife blender until they lost all graininess and became a lemon-yellow color. The white meat of the coconut was grated, 1 cup of boiling water was added, then it was mashed and kneaded for about 10 minutes, and finally put into a cloth and squeezed, yielding about ½ pint of thick coconut cream.

Since sautéed Bluegill fillets would break apart and become mush with all the stirring necessary to make a smooth Newburg, I was forced to adopt a different procedure. We used 6 of the little fillets per person, and an egg was beaten with a tablespoon of water and mixed thoroughly with the fish in a mixing bowl until each fillet was evenly dampened. Then, after leaving the fillets awhile in a strainer, so the surplus egg could drain away, we shook them in a bag with fine cracker crumbs until they had accumulated a good even coating. They were then fried in bland oil to a light golden brown, placed between paper towels to drain off any grease, and kept in a low oven so they would stay warm. The corn crêpes were finished and kept in the same oven while the sauce was made.

To make this sauce I melted ¼ cup of butter in the top of the double boiler and laced it with ¼ cup of cooking sherry. The coconut cream was merely stirred into the fish roe, for I was afraid that any attempt to beat it with a mixer would only result in coconut butter. This cream-roe mixture was added to the butter and wine, and the stirring began. As I stirred constantly I used the other hand to add 1 teaspoon salt, a dash of black pepper, and a pinch of nutmeg. The second it began to thicken I removed it from the boiling water, but kept stirring another minute, so there would be no lumping about the edges.

We had 6 large, but very thin, corn crêpes, so I put 3 of the fried fillets on each cake, added a generous spoonful of the sauce, and rolled it up. The remainder of the sauce was poured over the rolls, leaving one dry spot on each for the cognac, for these were served complete with the flaming brandy.

I served this with a wild-watercress salad and wild-strawberry short-cake for dessert, and we envied no man in the world his dinner that evening.

10. Soft-Shell Clams, Long-Necks, or Steamers

(*Mya arenaria*)

ALSO called Long Clams and Nannynoses, this is *the* clam of New England, and a highly valued food clam throughout its range, which extends from Labrador to North Carolina. Some seventy-five years ago, it was accidentally introduced on the West Coast, where it thrived; today it is the largest-selling clam in California markets, and it has become abundantly naturalized on all suitable tide flats from Monterey to Alaska. Next to the oyster, it is now the most important food mollusk on both coasts. In many places on both coasts it is "farmed" or cultivated like the Oyster.

New England and the Maritime Provinces of Canada, with their deeply indented and island-studded coasts, have literally thousands of tidal flats where this clam can be found in great abundance twice each day at low tide. Many millions of pounds are dug and marketed each year, and more millions of pounds are dug and eaten by non-professional diggers and hence do not appear in the market figures. It is probably the food mollusk that is most available to the camper and forager both in New England and on the West Coast. In the mid-Atlantic states it is not so often found in the intertidal zone, and most of the clams in this area are taken with dredges and tongs.

The Steamer gets to be 4 or 5 inches long if it is lucky, but persistent digging has made clams of this size rare, and most specimens found will be from 2 to 3 inches or smaller. The shell is moderately thick and gapes at both ends, that is, it cannot be completely closed. Despite its thickness, this shell is very brittle and easily broken, and the big problem in digging Steamers is to get the clam to the surface without cracking the shell; otherwise, sand gets into the clam's mantle cavity and is almost impossible to remove. It is naturally a dull white, or chalky in color, but may be stained a darker color by the ground it lies in. The shell is oval in outline, and the surface is roughened and wrinkled by lines of growth. It has very long siphons united throughout their length, and this compound siphon is called the clam's "neck."

The Steamer is regularly found on tidal flats that are free of water twice a day. Good digging ground is located by walking across such flats and observing the size and number of squirts that occur near your feet as your footsteps disturb the clams. These squirts are caused when the clam suddenly withdraws its siphon as it senses the approach of danger. The size of the clams lying beneath can be judged somewhat by the size of the squirts and the size of the holes left when siphons have withdrawn, but I have dug beneath many patches of holes that appeared very promising to find some pretty small clams. However, these small clams are as good as or better than the large ones, although more tedious to remove from the shell. Formerly many states in the Steamer Clam area imposed a minimum size on the clams that could be taken, but on the advice of biologists and conservationists the present tendency is to abandon such limitations. Clam diggers measure their catch by the peck, bushel, or basketful and not by number. This would seem to make a minimum-size limit desirable, as each basketful would then contain fewer clams, but this is not the way it works. In practice the digger would discard all undersized clams, which are promptly eaten by the sea gulls, and consequently would have to dig up a larger area in order to fill his basket. Between the digger and the sea gulls a great many more clams are destroyed when there is a minimum-size limit than when there is not.

When a likely spot is located, the clams are most easily removed by digging a hole near the center of the area and working out from this, turning each new forkful of sand back into the hole. In Maine, a stout, short-handled fork with the tines at right angles to the handle is almost invariably used for the digging, but an ordinary garden spading fork will do, especially if two people are working together, one turning over the sand and another picking up the clams. In a good spot from one to four bushels can often be gathered at one low tide, but of course the average camper or forager will not want such a large quantity at one time, nor will the laws of most states allow him to take so many.

Freshly dug Steamers are apt to be gritty or sandy. There are two very good ways to rid them of the contained dirt. One very good way is to put them on a rack in a large kettle or tub, cover them with fresh water, and sprinkle a small handful of cornmeal in the water. Change the water twice a day, adding a little new cornmeal each time, and soak the clams two days before shucking and eating. Steamers can be treated this way without killing them, although such long soaking in fresh water and close quarters would be fatal to most other clams, and indeed to nearly all other forms of seashore life. Steamers naturally tolerate considerable fresh water and are often found on the shores of tidal rivers so far from the sea that the water is merely brackish. They can stand being soaked in close quarters in still water because they have the unique ability to live for a considerable time with no oxygen at all. A scientist once kept some of these clams for eight days in a completely oxygen-free environment, and they were alive and healthy after the ordeal. There are many anaerobic bacteria that not only tolerate but require an oxygen-free medium in which to develop, but it is pretty startling to discover the same ability in a creature as highly organized as a mollusk. These airless clams didn't go into any kind of suspended animation, and all life-processes continued to operate. They even continued to liberate carbon dioxide as a waste product, so obviously they were getting oxygen from somewhere. The only change in the clams that was noticed during this time was a gradual decrease in the stored glycogen (a kind of animal starch or sugar) in the clam's "liver." By some process that is still unknown, this clam apparently is able to synthesize oxygen from glycogen.

The second way to rid these clams of mud and sand is to use the same technique that I use for de-sanding and de-mudding the *Macoma* Clams of the West Coast: this is to place the clams in a tightly covered basket made of galvanized-wire hardware cloth and suspend them from the end of a pier or wharf in clean seawater for two to three days. I recently tried this method on Steamers in Maine, and it works to perfection. There are such tremendous tides in that part of Maine that I had difficulty finding a place where I could suspend my clams in water where they would not be hanging in the air at low tide. I finally found a floating dock that had six feet of water under it at extremely low tide and hung my clams just clear of the bottom when the tide was out. At the next low tide I went down to see how my clams were doing and carried my fishing rod along to see how fishing was from this floating dock. Right here I made an unexpected and important discovery. It seemed that this wire basketful of inaccessible clams was proving a great attraction to the fish population. I baited with a small clam, and by fishing near where the basket was hanging I began pulling in fish one after another. I caught Flounder, Cunner, Tautog, Pollack, and mackerel, and had a great string of fine fish within an hour. During the rest of my stay there I kept a basket of clams hanging below that dock at all times and had a constant supply of both clams and fish.

The Steamer is the clam that has been made famous by the renowned New England Clambake, and a complete account of such an outing with directions for preparing the bake will be found on pp. 92-97. However, it is not absolutely necessary that a Clambake be prepared on the beach, nor is it essential that it be prepared for a large crowd. While staying at this same place in Maine, my wife and I invited a couple in the cottage next to ours for dinner one night, and we prepared a perfectly wonderful Clambake for only four people, right on the small gas stove in our rented cottage.

On the day we gave the dinner, there was an extremely low tide at midmorning. The water was calm and perfectly clear, so by closely examining rocks, clumps of seaweed, and patches of eelgrass just before the tide reached its lowest ebb, I found a dozen fine, large Jonah Crabs, which I captured with a dip net. These I confined in the wire basket with my clams, to keep them fresh and alive until evening. Then I baited a hook with clams and caught a few fresh Flounders, Pollack, and one mackerel, before the tide came in. I also gathered a half-bushel basketful of fresh, live Rockweed, *Fucus*, from the rocky shore. I filleted the fish and cut the flesh into boneless, finger-sized pieces.

I always carry a large canning kettle on my vacation trips, for I invariably return with many jars of canned wild berries and other wild fruit, pickled fish, jams, jellies, and other delicacies. I almost make my vacations pay. That evening we put this large kettle with about an inch of water in the bottom of it on the little gas stove, and when it boiled we added a 6-inch layer of Rockweed. Then we scrubbed some of those beautiful Maine baking potatoes, pricked the skins with a fork to keep them from bursting, bedded them in the Rockweed and covered them with more Rockweed. As soon as it was steaming freely, we covered it tightly and lowered the heat a bit. When the potatoes had steamed for 30 minutes, we added all the de-sanded clams that four of us thought we could eat, the Jonah Crabs I had captured that morning, some fresh-plucked ears of sweet corn bought from a nearby farmer and the fish fillets, each piece wrapped in a corn husk. The outer husks of the corn had been removed and used to wrap the fish, and the inner husks had been opened to make sure we would cook no lurking worms and to remove most of the silk from the sweet corn. After all the food was in, the kettle was filled to the top with wet Rockweed, covered tightly, and allowed to steam for another 20 minutes. Rockweed is used in such a Clambake, not only because it is a convenient and easily obtained steaming medium, but also because of the flavor it contributes to the food that is cooked in it. Chew on a piece of Rockweed and you will see what I mean. At first it seems almost tasteless, then as you continue chewing you detect a very sweet, and at the same time savory, flavor. It is neither tender enough nor good enough to be considered really edible, but the flavor it can transfer to other seafoods with its sweet steam is completely delicious.

When the steaming was finished, I lifted out the Rockweed with a pair of kitchen tongs and heaped everyone's plate with the fragrant food, then replaced the hot weed to keep the seconds and thirds warm. We had a bowl of Melted-Butter Sauce on the table, some garlic butter, a big container to receive the clam shells and crab shells, and I broke out a bottle of dry Elderblow wine * that I had made three years before. We all ate until we could hold no more and we all agreed that we had never tasted a better Clambake.

For really precise timing in your cooking, this kind of indoor

* For complete directions on how to make your own Elderblow Wine, see pp. 94-95 of Stalking the Wild Asparagus.

Clambake is better than the outdoor variety. All the food must go into the steaming-pit at the same time when you prepare a Clambake on the beach. Actually, though, potatoes require 50 minutes' steaming to cook through, while clams, crabs, Lobsters, or Mussels require only 20 minutes to be brought to perfection. Sweet corn and small sections of fish fillets wrapped in corn husk are both better steamed only about 15 minutes. Indoors, you can add each food to the steaming-kettle according to its required cooking time and get more gourmet results.

When Steamers have been thoroughly de-sanded, they can be used in almost any clam recipe found in this book. A great favorite at the Maine drive-in restaurants is French Fried Battered Clams. Open the clams over a mixing bowl to catch the liquor. Steamers are very easily opened, for one needs merely to slip a thin knife into either of the gaping ends of the shell and sever the adductor muscles. Cut the clam loose from the shell and snip off the dark-colored "neck" with a pair of kitchen shears. A pint of drained clams will serve four people generously. To make the batter, mix 2 well-beaten eggs with ½ cup of clam liquor, 1 teaspoon salt, and ¼ teaspoon black pepper. Add this mixture to 1 cup flour that has been sifted with ½ teaspoon baking powder and stir into a smooth batter, which should be the consistency of very thick cream. Dip the clams, one by one, in the batter, and drop into fat that has been heated to 375°. Remove the clams when they become golden brown, and drain on paper towels. Serve while still hot.

For perfect Clam Newburg, make it in the top of a double boiler. First put the top of the double boiler over direct heat, melt 1 tablespoon butter in it, then add 1 tablespoon flour and stir to a smooth blend. When it starts to brown, put over boiling water, add 1 cup coffee cream and from 24 to 36 shucked, drained Steamers, the number depending on their size. Be sure to snip the black "necks" from the clams before adding them to this dish. Cook over boiling water for 10 minutes, add 2 well-beaten egg yolks, then cook and stir for 2 minutes more. Add ½ teaspoon salt, a tiny pinch of cayenne, and 2 tablespoons cooking sherry. Stir once more, then serve over toast rounds or over Corn-Cream Cakes (page 99). A bottle of Moselle, cooled to just 50°, will be exactly right to serve with this dish.

11. The Neglected Blue Mussel:

A SAVORY SEAFOOD TREAT

(*Mytilus edulis*)

ONE of the most delicious, most abundant, and most easily procured of all seafoods is the Blue Mussel, so common on our shores from the Arctic to Cape Hatteras. Unfortunately and unexplainably, this tremendous resource of excellent food is almost totally ignored in this country. How different things are in France—that country of wonderful food and fine cooks. There the mussel is considered a luxury, and like the Oyster is cultivated in artificial beds, because the natural production of mussels along the French coast would never equal the demand. This is exactly the same species that is spurned on our own coasts, or at best only eaten by recent immigrants who still remember its worth. Its scientific name means "edible mussel."

If mussels ever become rare or extinct, it will not be because of any lack of fertility on their part. A single female mussel may release as many as fifteen million ova at a single spawning, and the mussel spawns a number of times each year, between April and October. Some sea change imperceptible to us gives the signal, and every mussel in the area spawns at the same time. In New England there are mussel beds of vast extent, containing literally billions of individuals, and when every female starts shedding millions of eggs at the same time, the sea becomes clouded with mussel ova. One would think that very few of these eggs ever achieve fertilization, since they are merely entrusted to the tides and have to meet a male sperm in the mighty ocean, but this, apparently, is not a problem. Despite the uncounted trillions of eggs released at the same time, the male mussel seems to be equal to his task. Place a drop of sea water under a powerful microscope, and you may see dozens of mussel ova, but around each egg will be a great swarm of excited sperm trying to penetrate the outer membrane. Only one ever succeeds, for as soon as one sperm manages to enter the egg, the outer ovarian membrane undergoes what seems like a magical change and becomes impenetrable.

All this mad, molluscan love life naturally results in plenty of mussels. Toward the southern limit of their range, one must search for this neglected delicacy, pulling them from pilings near the low-water line, but in many places along the rocky shores of New England one merely has to walk out at low tide and select the best ones from an unlimited supply. Adult mussels are about 3 inches long; the shell is blue-black outside and deep violet inside. Mussels spin very strong silken black threads, called the byssus, or beard, with which they attach themselves to their anchorage. An unattached mussel is probably a dead one; don't take it. As I gather each mussel, I grasp the protruding threads of the byssus between thumb and forefinger and pull toward the large end, debearding it on the spot.

The mussel can hold its closely fitting shells so firmly together that it is very hard to insert a clam knife, but this never bothers me. I care nothing for raw mussels, so I circumvent its defenses by steaming it open.

If mussels are to be the main dish you will need from one to two dozen for each hearty eater. The first step in all mussel cookery is to scrub the outside of the shells thoroughly. I use a scratch pad of the kind intended for cleaning dirty kettles on Blue Mussels and a very stiff brush on the Ribbed Mussels.

For Plain Steamed Mussels, which is not a dish to be despised, place 4 dozen mussels in a deep kettle, add ½ cup of dry white wine, and steam for 20 minutes, when you will find all the mussels gaping wide. The liquid that runs out of the mussels as they cook will have added materially to the broth. Remove the top shells and

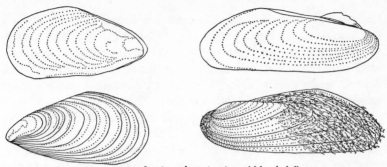

Interior and exterior views of Mussel shells.

Left, Mytilus californianus, about 3 inches long.   Right, Modiolus rectus, about 4 inches long.

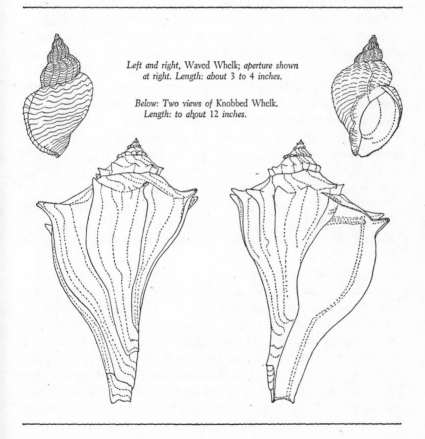

*Left and right, Waved Whelk; aperture shown
at right. Length: about 3 to 4 inches.*

*Below: Two views of Knobbed Whelk.
Length: to about 12 inches.*

serve hot on the half shell, with the broth poured over the mussels and a flavorsome Melted-Butter Sauce (page 21) on the side. Even the most rigid teetotaler need not bridle at the use of wine in seafood cookery. It is strictly for flavor, and not for wassailing. When wine is cooked, all the alcohol evaporates, but it imparts a flavor impossible to achieve any other way.

If you wish to be a little more elegant with Steamed Mussels make a roux to replace the more plebeian broth and melted butter. Beat together 2 egg yolks and ¼ cup of rich cream, then add 1 cup of the winy broth from the steamed mussels. Cook and stir in the top of a double boiler until the roux thickens to the consistency of smooth gravy. Add chopped parsley and chives, pour this over the mussels on the half shell, and serve hot.

Fried Mussels are good in camp or at home. Scrub the shells of 4 dozen mussels, place them in a deep kettle with ½ cup water, and steam until they open wide. Remove the meat and discard the shells. Beat 1 egg slightly with 1 tablespoon of water. Stir the mussel meats in the egg until each mussel is evenly coated, then allow the mussels to sit in a coarse strainer for 5 minutes so the surplus egg can drain away. Put 1 cup of fine cracker crumbs in a stout paper bag, dump in the mussels, and shake them until they are evenly coated and nicely separated. Fry in 375° fat until they are a light golden brown, about 3 minutes. Drain on paper towels and serve, still piping hot, with a garnish of parsley and lemon wedges.

Contrary to popular opinion, fried foods need not be greasy foods. The mussel, besides being a veritable storehouse of many of the minerals and vitamins needed for perfect nutrition, contains some of the most perfect proteins in all nature. It is, however, deficient in fats and carbohydrates, and despite all the preaching of food faddists, and diet addicts, the human body needs fats and carbohydrates for proper nutrition. A mussel, coated with egg to act as a fat barrier, rolled in cracker crumbs to furnish carbohydrates as well as interesting texture, cooked in fat at just the right temperature, and then quickly drained on paper towels, is not nearly as greasy as one merely steamed and then doused with butter. Fried Mussels are not only delicious, they are a complete food, and they will never cause you to put on more weight than you will take off by the healthful outdoor exercise of gathering them from the rocky beach.

Another excellent camp recipe is Mussels with Scrambled Eggs. Steam open the mussels, remove them, and chop coarsely. Allow 6 mussels per egg and 2 eggs per person. Melt some butter in a heavy frying pan over medium-low heat, and scramble the mussels and eggs together just as you would scramble eggs alone. Serve with buttered toast and hot coffee. If available, a tablespoon of chopped chives for each serving, scrambled right with the eggs and mussels, adds a wonderful savor. This may sound like a breakfast dish, but I have found that it also makes a welcome main dish for a camp luncheon or dinner.

The French like to cook the seasoning materials into mussels as they steam them. My own adaptation of a recipe for French Mussels will feed four people at home or two in camp, where appetites tend to be uncontrolled. Scrub 4 dozen mussels and put them in a kettle with ½ cup of dry white wine, 2 chopped scallions, 2 sprigs of fresh thyme or ½ teaspoon powdered thyme, and 6 dried Bayberry leaves. Steam 20 minutes, then remove top shells and beards, and arrange the mussels on the half shell on a large platter. In a heavy frying pan melt 3 tablespoons of butter, then stir 1½ tablespoons of flour into it to make a paste. Let this cook 5 minutes, without browning, then add 1½ cups of strained broth from the steaming-pot, and cook and stir until you have a smooth, thin gravy. Pour this over the cooked mussels and serve with thick, crusty slices of French bread. This is no dish for table niceties. To enjoy this savory gravy, you simply must sop your bread in it.

The only recipes I could find for Stuffed Mussels called for a thick bread stuffing over whole mussels in the half shell. I didn't like having to eat several bites of pure stuffing in order to get at the mussel, so I devised my own double-rich recipe. Scrub and steam open 4 dozen mussels, remove the meat, and chop it coarsely. Chop 1 medium-sized onion and sauté in butter to a clear yellow color, then add 1 tablespoon of chopped parsley, ¼ teaspoon poultry seasoning, and 1½ cups of bread crumbs. Stir and cook for about 2 minutes, then remove from the fire and stir in the chopped mussels and enough of the broth from the steaming-kettle to dampen slightly. Mix well, then fill 2 dozen of the half-shells with this mixture. This means there are two whole mussels in each half-shell, so you're really eating mussels, not just seasoned bread crumbs. Arrange the stuffed half shells on a cookie sheet, place under the broiler for a few minutes, just until the tops are light brown, then serve sizzling hot.

For a really hearty meal with an Italian flavor, try Red Mussel Spaghetti. Scrub, steam, and remove the shells from 4 dozen mussels. Chop the meat coarsely. Finely chop one large onion and one clove of garlic, then sauté them in 2 tablespoons olive oil until they are yellow and clear. Add ¼ cup finely chopped celery heart, using leaves and all, 1 teaspoon chili powder, one 10½-ounce can strained tomatoes, and one 6-ounce can tomato paste. Pour in the broth from the steamed mussels and simmer for one hour. Add 1 tablespoon chopped parsley, a sprig of thyme, 2 tablespoons butter, and the chopped mussels. Simmer for 10 minutes more, then serve over cooked, unsalted spaghetti.

Those who live outside the range of *Mytilus edulis* may be tempted to try the Atlantic Striated Mussel, *Modiolus demissus*, in some of these recipes. This handsome species with a ribbed shell, iridescent inside, is 3 to 4 inches long and is very common in tidal streams, salt marshes, and brackish bays, from Nova Scotia to Florida. In color it is yellowish green or bluish green on the outside, with rubbed spots showing the pearly iridescence of the interior.

The literature on the Striated Mussel is very cautious about its culinary qualities, one book saying, "usually considered inedible," and another even hinting that it might be poisonous. I have eaten Striated Mussels on several occasions and suffered no ill effects from it. In quality it varies considerably, according to the type of mud on which it grows, but at its very best it is not as good as the Blue Mussel at its worst. The Striated Mussel may be wholesome enough, but it simply is not as good as the Blue Mussel.

The Bent Mussel, or Hooked Mussel, *Mytilus recurvus*, is a small species found from New Jersey to Florida. The shell is triangular, about 1½ inches long, and strongly curved near the apex. I have tried eating this mussel with some success, but, while it will stave off the pangs of hunger, it will not furnish the taste treat that is offered by the Northern Blue Mussel. I found them barely edible when covered with a tasty sauce, but with none of the superb quality found in their Northern relatives. You will find them in great clusters attached to mangrove roots, rocks, and old oyster shells.

Is there danger of being poisoned by eating some of these little-known species of shellfish? Not if a few simple and easily learned precautions are taken. During the summer and autumn, the mussels along the Pacific Coast may ingest certain poisonous, microscopic plants along with the other food they strain from sea water. This poison is stored in the creature's "liver" and may reach concentrations that are poisonous to man. Mussels are quarantined along the California coast from May to October. This practice is not new. The saltwater tribes of Indians who inhabited the Northwest Coast before the coming of the white man well knew of this danger. They dearly loved seafood, indeed they subsisted almost wholly on the products of the sea and the beaches. Through generations they accumulated much sea-lore and learned to read the signs of the seasons and the sea. The dinoflagellate, *Gonyaulax*, the genus that causes most of this trouble, is luminescent, and when it multiplies into dangerous numbers it brings a recognizable glow to the sea. Then every wave carries a crest of fire; the wake of a boat looks like a moon-path across the sea, while schools of herring and candlefish become displays of shooting stars. Beautiful, but ominous. The Indians knew that mussels were not to be eaten while the sea was glowing. They humanely posted guards along the trails to warn inland Indians, who could not read the signs of the sea as well as their saltwater brethren, that they were not to eat mussels until the sea had again become dark and wintry. During the mild, open winters of the West Coast, the mussel again becomes wholesome and delicious.

Another danger faces those who would gather their own food from the seashores. While the mollusks mentioned in this book are all wholesome in themselves, because of their feeding habits, the bivalves are peculiarly apt to pick up pollution of human origin and pass it back to the same species that was responsible for its being in the water in the first place. These creatures continuously draw seawater through their siphons and strain everything edible from it, and some of the things a bivalve might consider edible would horrify you. This applies to the commonly accepted oysters and clams even more than to the mussels, for oysters and clams are often eaten raw, while mussels are only palatable when cooked and consequently thoroughly pasteurized.

One should never eat any kind of shellfish that is not perfectly fresh. A mollusk that is dead and beginning to spoil is dangerous, but this is also true of many other kinds of food. The poisons that develop in tainted foods, popularly known as the ptomaines, are the result of putrefaction, and shellfish have no monopoly on such toxic substances. To be perfectly sure of avoiding this kind of trouble, see that the mollusk you intend to eat is not only fresh, but actually alive when you start to prepare it. A raw mollusk that doesn't resist your efforts to open its shell should be discarded.

The mollusks mentioned in this book do not generate any poisonous substances within themselves. This is not to say that all mollusks found on our shores are edible, although most of them are. Some, like the Striated Mussels, are so disagreeable in taste, texture, or odor that one would not find them attractive as food. Others, like the giant Horse Conch of the Florida shores, are so peppery-hot as to cause discomfort to anyone even tasting them, but even these, although not considered edible, contain no dangerous poisons and would furnish nutriment in emergencies.

The West Coast is well supplied with edible mussels, but they can only be eaten from November through April. For the other five months, Pacific mussels feed on certain plankton that makes them slightly poisonous to human beings. But during the mild and open winters of the West Coast, the mussels are excellent and in many places abundant.

Our old friend, *Mytilus edulis*, or the common Blue Mussel, has been introduced on the West Coast and is becoming abundant in some areas. Where found, it is just as good as when it grows in the East, if gathered during the winter months.

The *Mytilus californianus* is about the same size as the Blue Mussel, and very like that species, except that its shell is a light brown. The mussel is likely to be orange-colored inside, due to an abundance of highly-colored ova that seem to fill the whole animal. Far from detracting from its edible qualities, this bright-colored gonad only adds to the attractiveness and flavor of these mussels.

The *Modiolus rectus* has a narrow shell, up to 4 inches in length, with a glossy, dark-brown epidermis, and a white interior. Both of these latter species are found from Vancouver, B.C., to Lower California. I have enjoyed many a meal from *M. rectus* gathered from rocks on the shores of Puget Sound when the tide was extra-low.

The mussel family probably constitutes the greatest unused seafood resource to be found along the coasts of America. They are not only perfectly edible, but so delicious that many European epicures prefer them to any other molluscan seafood. Let's forget the senseless prejudice that has kept us from enjoying this fine food, and help ourselves to health and some mighty good eating that is waiting for us by the thousands of tons along our seashores.

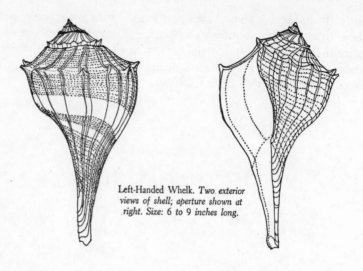

Left-Handed Whelk. *Two exterior views of shell; aperture shown at right. Size: 6 to 9 inches long.*

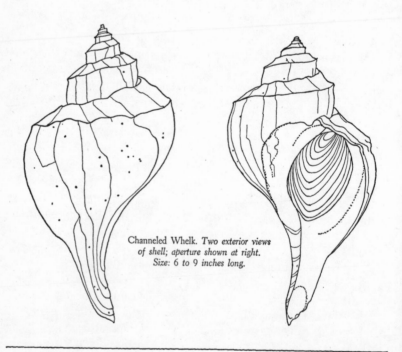

Channeled Whelk. *Two exterior views of shell; aperture shown at right. Size: 6 to 9 inches long.*

12. Moon Shells and Left-Handed Whelks:
THE LARGE EDIBLE GASTROPODS

AMERICA is blessed with many kinds of large edible gastropods that are easily available to the beachcomber. The fact that many Americans reject these perfectly delicious seafoods, because of an entirely illogical and unreasonable prejudice, should not bother us, indeed we should be glad such prejudice exists, for when everyone discovers what a good thing he has been missing, these creatures will become rare and expensive instead of being available in their present abundance. These seafoods are eagerly sought and highly appreciated in many countries, and immigrants from those countries long ago discovered that our American gastropods were just as delicious as the ones they had left behind. Whelks sometimes appear on big city markets in this country, but they are sold only to those discriminating immigrants from countries where the large, snail-like creatures are eaten.

It is hard to understand this American prejudice against eating snail-like mollusks, for this prejudice did not exist in any of the countries from which the majority of Americans came. Nor did the early settlers find it among the Indians over here, for these aborigines freely ate all these creatures and in some instances seemed to prefer them to other kinds, as the great shell mounds or "kitchen middens" they left behind show. Even in modern America this prejudice is inconsistent, for the conch is widely eaten in the South, while on the West Coast the Abalone is considered a delicacy and sold at luxury prices, and both these creatures have single, coiled shells and are definitely within the snail-like division of gastropods. If you are infected with this prejudice, the easiest way to cure it is to sample a well-cooked dish made of the flesh of one of these creatures.

To introduce this group let us first discuss the Waved Whelk, *Buccinum undatum*, the common Edible Whelk of the British Isles and northern Europe. It is equally plentiful on this side of the Atlantic, but is almost unknown in American markets, although our local Waved Whelks are identical in flavor, texture, and general edibility with those sold by the hundreds of tons in the British Isles.

A good idea of the over-all appearance of the Waved Whelk can be gained by studying the drawing on page 111. An average specimen is 3 to 4 inches long with a comparatively high spire made up of 6 to 8 whorls, but the lower, body whorl makes up considerably more than half the bulk of the animal. The spire is sculptured with longitudinal ridges or wrinkles, somewhat uneven in size and placement, but these disappear as they approach the lower part of the body whorl. These are the waves or undulations that give this Whelk both its common and its scientific specific names. The outside of the shell is rusty brown or gray in color, and the inside is white or yellow with a smooth porcelain-like finish. On the back side of the large foot there is a horny operculum, or trapdoor, that neatly closes the opening when the creature completely withdraws into its shell. This Whelk is fiercely carnivorous, eating any animal flesh, dead or alive, that it can find.

The Waved Whelk lays its eggs in yellowish capsules about the size and shape of split peas, and each capsule contains hundreds of eggs. The capsules are attached to one another in an overlapping manner making up masses as large as your fist, and these egg masses are often found stranded on the beach in areas where Waved Whelks are abundant. If these egg capsules are crushed and rubbed with water they will lather like soap, and fishermen and sailors often use them in washing their hands, calling them "sea-wash balls." In Europe, Waved Whelks are found from Iceland to the Mediterranean coast of Spain; on this side of the Atlantic they are found from Newfoundland to Cape Hatteras, but they are only abundant in the northern part of their range.

I had long been familiar with the shell of the Waved Whelk, but my first encounter with the living animal was near a little fishing village in Maine. I was on a floating dock, fishing for crabs with some ring nets that I had made by lacing fish netting inside wire barrel hoops and attaching bridles and lifting lines of stout twine. For bait I tied an old fish head in the center of each net. So far as I could discover, this method of crabbing had never before been used in that area, and the local boys gathered around and gazed in wonder as I began pulling in some very nice Rock Crabs and Jonah Crabs. Those high fish heads, however, began attracting Waved Whelks, and soon I was catching more whelks than crabs.

A friendly lobsterman who was docking his boat happened to notice my growing hoard of Waved Whelks. "Do you *want* them things?" he asked in astonishment. I explained that I was interested in all sizable mollusks, and he said, "I'll give you alla them things you want. They always get in my lobster traps and steal the bait." I pointed out the cottage where I was staying, and that evening he came to the door with a large pailful of "them things." After that, every time he ran his lobster traps he would appear with a huge pailful of whelks. I like whelks; they are good food, but I'm not crazy enough about them to live on a whelk diet exclusively, so I was hard pressed to use this constant and embarrassingly large supply. In self-defense I developed the pickle recipe given below, and came home from my vacation that year with a generous supply of Pickled Whelks.

For my first experiment with cooking Waved Whelks, I simply boiled some in seawater, then slipped the creatures from their shells with a nutpick. The large, fleshy foot makes up most of the bulk of the creature, and this is the edible part. It is a cylinder about 1 inch in diameter and about 1½ inches long, white or grayish in color when boiled, but unfortunately it is streaked or spotted with black, which somewhat spoils its otherwise appetizing appearance. Boiled Whelk has a sweet, seafood taste, and when eaten hot with melted butter it is good, hearty food, though somewhat rubbery in texture. However, boiling it isn't the very best way to prepare this creature for the table.

To make Pickled Whelk, I boiled a pailful of the whelks in heavily salted water and slipped them from the shells. The viscera above the fleshy foot was discarded, and the operculum was pulled from the foot. In the bottom of each quart jar I put one bag of crab-boil spice and three Bayberry leaves. The jars were then loosely packed with cleaned whelks and finely sliced onions in alternating layers and the whole covered with boiling vinegar, then sealed. Shortly after returning home, we opened the first jar and found them delicious. Sliced into thin disks and speared on toothpicks, these savory Pickled Whelks are now our favorite appetizers, hors d'oeuvres, or cocktail snacks.

To prepare Waved Whelks for eating in the fresh state, it is better not to boil them first; remove them from the shells raw. To do this it is necessary to break the shell. Lay the shell on a firm surface and strike it a smart blow with a hammer, and it will break as easily as a piece of your best china. Remove the fleshy foot, discard the viscera and the broken shell, and remove the operculum by slipping a knife blade under it and pulling it off. Grind the feet in a food chopper using a medium-fine plate, and with each pint of whelks also grind 1 medium onion, peeled and diced, and about 1 ounce of cooked ham, fat and all. Work the ground mixture with your fingers until it assumes a spongy texture, then shape into patties or small balls with dampened hands. These Whelk Patties can be fried in bacon drippings or butter-flavored fat until they are nicely browned, then served hot. The little balls can be dropped into deep fat and French fried, or they can be dropped into soup about 5 minutes before it is to be removed from the fire.

I use this whelk mixture when making Deviled Whelks but I do a bit of nature-faking by cooking it in scallop shells or clamshells. It is amazing how much more acceptable a whelk becomes when camouflaged in this manner. To a pint of ground whelk mixture I add 2 beaten eggs, ½ cup bread crumbs, and ¼ teaspoon black pepper. It should be salted to taste, as whelks vary in saltiness according to the brininess of the water from which they were taken. Stuff the shells with the whelk mixture, and either dot the top with butter or stretch a piece of bacon over each mound. Bake 20 minutes in a 375° oven, or until the bacon is nicely cooked. Serve while still sizzling hot.

A friend of mine who lives in New Jersey recently took a vacation in the Bahamas, and on his return he was telling me of the delicious conch chowder, conch salads, and fried conchs that he had enjoyed there. He was amazed when I told him there were two very common edible mollusks that could be caught along the New Jersey shore that could be used in making these same dishes, and they would be as delicious as any the West Indies ever produced. These are the Giant Whelks, which are very common along sandy shores from southern New England to northern Florida. They are the largest mollusks on the East Coast north of Cape Hatteras.

The larger of these two is the Knobbed Whelk, *Busycon carica*, and it is a great, pear-shaped snail that, despite its low twisted spire, sometimes reaches a length of 12 inches. There are 6 whorls and on the very large body whorl, just above the elongate oval opening, there is a ring of heavy, blunt knobs, making the whole shell look something like the caveman's club of the cartoonists. On the forward end (the end opposite the twisted spire), the shell is reduced to a slender open canal. It is gray in color, but the lining of the aperture is a dull brick-red. The shape of the shell can be seen from the drawing on page 111.

The Knobbed Whelk has a large, fleshy body and a very broad foot upon which it crawls, carrying the shell with the spire backwards and the canal forwards. In the canal is the incurrent siphon tube. Just under the siphon is the head, with a pair of stout tentacles, each with a fairly well-developed eye on its lower, outer edge. From beneath the head a long proboscis, like an elephant's trunk, extends forward, terminating in a mouth containing a rasping, file-like tongue that is called a "radula." When feeding, the Knobbed Whelk grasps a clam, mussel, or oyster with its stout foot and strikes it repeated blows with the heavy knobs on its body whorl until the bivalve's shell is broken. Since the whelk lives on such succulent food, it is no wonder that its own flesh is sweet and delicious.

The eggs of the Knobbed Whelk are laid in double-edged, parchmentlike, and disk-shaped brown capsules about 1 inch in diameter, and these capsules are held together near one edge by a cord of the same parchmentlike material. Each string may consist of more than a hundred capsules and be a yard in length. These egg-strings are often cast ashore, causing much wonder among casual beachcombers about the nature of the creature that produced them.

Both this and the following species can be prepared in any of the ways recommended for the Waved Whelk, except that if you make Pickled Whelk the flesh of these large species should be diced to approximately bite sizes before pickling. In addition, the large foot and body of either of these large species can be sliced into Whelk Steaks. Make the slices about ½ inch thick; put each steak between two pieces of clean muslin and pound it thoroughly with a meat hammer, wooden mallet, or even with the rolling pin. Season with black pepper and a very little salt, roll it in flour, then fry to a light brown. Don't let the yellow color of the whelk's large foot put you off, for this color interferes with the sweet, fine flavor not at all.

The Channeled Whelk, *Busycon canaliculatum*, occasionally produces specimens as large as the largest of the Knobbed Whelks, but the average size is from 6 to 9 inches long, making it a bit smaller than the Giant described above. It is distinguished by having a deep, channeled groove following the whorls about at the suture. The spire is low, made up of 5 to 6 whorls that are quite square at the shoulders. The big body whorl makes up at least two thirds of the total length of the shell, and it is prolonged forward into a narrow, nearly straight, tubular canal. The shell is yellowish gray outside, and the interior is lined with yellow. In life the shell is covered with a brownish periostracum, and the outside of the shell is marked with many revolving lines. The egg-string of this species is very like that of the Knobbed Whelk, except that the individual capsules have single sharp edges rather than double ones. In range, habits, and habitat, they are much alike, and both are equally edible. Both these whelks sometimes appear in the Boston, New York, and Philadelphia markets, where they are bought by immigrants from countries where they do not have a prejudice against eating snail-shaped creatures.

Both kinds of Giant Whelks can be caught on ring nets similar to those I used to catch Waved Whelks. You are not likely to catch as many of the Giants, but then, one doesn't need so many when they're so large. I have caught them on fish-head bait while crabbing, but the best bait for these whelks is from the tough foot sections of Surf Clams. After every heavy storm, live Giant Whelks of both species get thrown on the shore, and then you can merely walk along the beach and pick them up. Be sure they're fresh and alive before you eat them. While the wild waves are still beating, one can often see them, sometimes only a few, sometimes hundreds, and sometimes thousands. At such times one can merely pick up all the whelks one can use, with no effort at all. Probably the surest method of all to catch enough of these big whelks for a meal is by skin diving. With scuba gear or even with an ordinary snorkel mask, one can bring them up from 2 or 3 fathoms of water, and this is about their favorite depth. Select your areas for whelk fishing by the number of shells that have been cast on the nearby shore, and you are almost certain of success.

South of Cape Hatteras the two species described above disappear, and in that area the most common whelk is the *Busycon contrarium,* the Left-Handed Whelk, or Lightning Whelk. At first glance the Left-Handed Whelk looks like a small and highly-colored Knobbed Whelk, for its shoulders are adorned with the same kind of knoblike projections. On closer inspection, however, one sees that the whole creature is built backward. When other species of whelk are held with the spire upward and the opening toward you, the opening is on the right-hand side of the shell, but with the perverse *contrarium* it is on the left, and the shell is twisted together in the opposite direction to that affected by other members of this genus. The color is striking, consisting of lightninglike streaks of rich brown radiating from the apex on a light fawn-colored background. This whelk has been known to attain a length of 12 inches, but the average specimen will be from 6 to 9 inches long. Their habits are very like those of their more Northern relatives, and they can be caught and cooked by exactly the same methods.

From Cape Hatteras to Florida, and around the Gulf Coast to Texas, one also finds the smaller *Busycon spiratum,* commonly called the Pear Conch. This species resembles the Channeled Whelk but is smaller, has more rounded shoulders, and is more graceful in appearance. From 3 to 5 inches long, the shell is flesh-colored, sometimes streaked or banded with brown. The Pear Conch is found in the same kinds of sandy habitats as its Northern kinsmen, but it prefers even shallower water, so it can often be picked up by wading out on sandy bottoms at low tide. Of course these little whelks will not furnish the huge gob of meat that is found in the Giant Whelks, but they are still good-sized shellfish when compared to most other edible mollusks, and a dozen of them will feed a family.

So many edible mollusks have limited ranges and can be found only in remote areas that it is a relief to come on a group that has representatives on both coasts from the Arctic to Central America. These are the large, globular Moon Shells which are not usually recognized as edible forms by snail-avoiding Americans, but which, nevertheless, can be made into noble dishes by knowledgeable cooks, and the raw materials for these dishes can be picked up from almost any shallow sandy bottom along the entire stretch of the Atlantic and Pacific shores of North America.

My first acquaintance with these large round shells was when I bent to pick up a smooth, globular, 4-inch shell from ankle-deep water in Puget Sound. I was surprised to find the shell attached to a huge foot, more than 8 inches across the bottom. The creature was obviously far too large for its shell. A 4-inch spherical shell will hold a lot of mollusk, but that tremendous foot contained at least four times the bulk of even that large shell. Suddenly, it turned on the sprinkler. There is no other way to describe it. From apertures around the bottom edge of that circular foot came streams of water exactly as if one had tipped up a sprinkling can. As it sprinkled, it shrank, and within one minute it had withdrawn into the shell and closed the opening with a horny operculum.

This was the Lewis Moon Shell, *Lunatia lewisi*, grayish-brown in color, sometimes attaining a 5-inch diameter. It is found on sandy bottoms in shallow water from British Columbia to Mexico, and its occurrence in large numbers in the shell mounds of the coastal Indians shows that it was a favorite food in ancient days. From California to Mexico, one finds the slightly smaller Recluz's Moon Shell, *Polinices reclusianus*, which is from 1 to 3 inches in diameter and has a semi-glossy shell, grayish in color, with brown or greenish stains. Its shell is heavier in proportion to its size than is that of the Lewis Moon, but for eating purposes the two are about equal in quality.

On the East Coast the finest Moon Shell is the Northern Moon, *Lunatia heros*, available from the Gaspé Peninsula to the Carolinas. It is found on compact sand flats just below low tide or deeper. It sometimes attains a diameter of 4½ inches, but most of those you will find in shallow water will be from 2 to 3 inches in diameter. The color is dirty white to brownish gray. The smaller Shark Eye, *Polinices duplicatus*, from 1 to 2½ inches in diameter, is found in similar situations all the way from Cape Cod to Texas. With the aid of the drawing and the above descriptions you will be able to recognize any of the species as Moon Shells, wherever found. For culinary purposes there is no need to distinguish between the species, as all are equally edible.

The egg cases of the Moon Shell are some of the most puzzling things found by the casual beachcomber in a stroll along the shore.

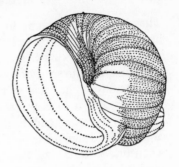

Northern Moon Shell. *Above, two exterior views; aperture or "mouth" shown at right. Size: about 2 to 3 inches in diameter.*

Left, Sand Collar, *or egg case of* Moon Shell.

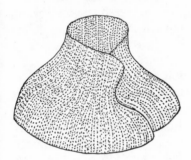

Green Sea Urchin. *Below, exterior view. Diameter: 2 to 2½ inches. Spines, about ½ inch long.*

Right, test (shell) *and anatomy, showing gonads.*

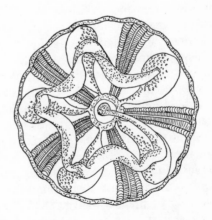

They look almost exactly like the rubber plungers used by plumbers in opening stopped-up toilets. The eggs of the Moon snail emerge from the mantle cavity in a continuous gelatinous sheet, and as this sheet emerges the snail turns about, wrapping it neatly about its huge foot. As fast as the sheet comes out it is covered with sand cemented together with a mucous secretion. These egg cases are commonly called sand collars, and an alternative common name for the Moon Shell is Sand Collar Snail. The sand collars are tough and rubbery when under water, but those cast ashore become brittle as they dry out and so fragile that it is almost impossible to carry one home without breaking it. Each sand collar contains up to 500,000 eggs all laid by a single snail. These hatch into tiny, free-swimming larvae that have a very high mortality rate during their first few weeks of existence.

The Moon Shells will eat any dead animal matter they can find, and this makes them vulnerable to the forager armed with ring nets baited with fish heads. I once left such a net down overnight, and the next morning when I pulled it up it contained eight Moon Shells. However, it is usually unnecessary to use ring nets to obtain enough of these mollusks for a meal. During very low tides, one can wade out on shallow, sandy bottoms and merely pick them up. I have even found them above the tide on the lower beach. Not all of them will have their shells showing, so look for small, round mounds of sand in which they hide. With a snorkel mask and a net bag tied to your bathing trunks to hold your catch, you can nearly always find a supply of Moons in shallow water at low tide.

The mighty foot of the Moon Shell can be ground with ham and onions as directed for the Giant Whelks, and the ground mixture can be used in the same ways. It is as Moon Steaks, however, that this food mollusk is really outstanding. Crack the shell away with a hammer, and the creature can be unwound from the central column. Discard the dark-colored viscera, slice off the operculum, and pick the meat clean of shell fragments. Slice it crosswise into steaks about ½ inch in thickness. Sprinkle each steak liberally with a commercial meat tenderizer, and then pound the meat thoroughly with a meat hammer or wooden mallet. Stack the steaks on a dish, alternating each steak with a layer of wax paper to keep them from sticking together, and set them in the refrigerator for 24 hours. After this aging time, dip each steak in egg beaten with a little cold water, then roll in fine bread crumbs and fry to a light brown. When prepared in this way these steaks are mild and tasty, and so tender they can be cut with a fork, but if you try to bypass the pounding and tenderizing and cook the huge strong foot of the Moon Shell by merely slicing and frying it, you will find it tough and rubbery. After Moon Steaks are covered with tenderizer and pounded, they can be placed in the freezer for future use, if you have found more Moons than you can consume immediately. Unlike most seafoods, which are not nearly as good frozen as they are fresh, the steaks of the Moon Shell actually seem to be improved by a month or more in the freezer.

13. Sea Urchins or Sea Eggs

W HO ever heard of eating a Sea Urchin? Well, at least several million people of the Christian Mediterranean countries have heard of it, for Sea Urchins appear regularly in their fish markets, and catching them is a valuable fishing industry in that part of the world. Chile, that narrow, South American country, with an extremely long coastline on the Pacific, has a tremendous variety of seafood at its doorstep, and yet the Chilenos consider the Sea Urchin one of the finest delicacies to come from the sea. The Italians who immigrated to California thought the local neglect of the excellent urchins that exist there a criminal waste of fine food, a situation they soon set about remedying.

The Sea Urchin has no flesh that I can discover, so only the roe, or gonads, are eaten, and they taste even more delicious and delicate than the finest caviar. I once offered some of this delicate roe to a cute young thing who was visiting us, but she found the thought of eating the eggs of Sea Urchins repulsive. I asked her why this food should repel her more than the sturgeon roe, which she admittedly loved when it was called "caviar," and she answered, "But Sea Urchins look so awful." Did you ever meet a sturgeon face to face? Talk about looking awful! Even a "virgin sturgeon," which is supposed to furnish the finest caviar, according to a bawdy song we used to sing in the Army, would never win a beauty contest. Actually, I consider the Sea Urchin quite pretty, and I think the dried tests—or shells—beautiful.

The Sea Urchin is akin to the starfish, but the relationship is not very apparent to amateur eyes until the urchin is dissected. Most species are globular in shape and covered with long, movable spines. These spines gave the creature its common name, "urchin," an old English name for the spiny hedgehog. Had it been named in America, it would have been called the Sea Porcupine, and the closely related Heart Urchin is sometimes called that. Its name in most European languages may refer to its delicious edibility as well as to its shape, for these names usually translate "Sea Egg."

Just under its spiny exterior, the Sea Urchin has a globular hard test, usually called a "shell" by laymen, made up of 5 plates of one design and 5 of another. It is this design on a pattern of fives that betrays this creature's relationship to the starfish. The mouth is in the center of the bottom side, and is actually 5-jawed. These 5 articulated jaws moving toward a common center are collectively called "Aristotle's lantern," because the whole apparatus is shaped like an antique, 5-sided lantern and was first described by Aristotle. Inside, the pentamerous design is continued in most organs, and just under the upper part of the test is the part in which we are interested, the 5-branched gonads, looking like a little starfish when removed entire and laid on a flat surface. These roe or sperm sacs range in color from almost white to bright orange, and the brighter-colored ones are considered best, but the gonads of all Sea Urchins found on American coasts are edible.

America is well supplied with Sea Urchins. While they may be absent from some stretches of shore because there is no suitable habitat for them there, there is no part of our seacoast that is not within the range of one or more Sea Urchin species. Generally, they will be found in depressions, crevices, and tide pools at about low-water line on rocky shores, often in bountiful abundance.

The most cosmopolitan member of the Sea Urchin family is the common Green Urchin that, all unknowingly, carries about the perfectly stupendous scientific name, *Strongylocentrotus drobachiensis!* Try that on your vocal cords. This is said to be the longest Latin binomial (or two-word name) in all zoology. The Green Urchin is found on our Atlantic coast from Greenland to New Jersey, and on the Pacific from the Arctic to Oregon. It is also found on northern Asiatic and European shores. The body is 2 to 2½ inches in diameter, the spines little over ½ inch in length, sharp-pointed and usually greenish in color, sometimes with a violet tinge. Or, the spines may be deep green and the test violet. Just to confuse things, one occasionally finds one that is dull brown all over. I have been places along the shores of Maine where one could have taken a bushel of these urchins from one tide pool or a barrelful from a few yards of shoreline, if anyone had use for so many.

Toward southern New England, the Green Urchin is gradually replaced by the Atlantic Purple Urchin, *Arbacia punctulata*, which ranges from Massachusetts Bay to the Gulf of Mexico. Slightly smaller than its Northern kinsman, this species has a test about 1½ to 2 inches in diameter, and although the spines average twice as long as those of the Green Urchin, these long spines are humanely rounded on the ends instead of being sharp-pointed. The color is usually purplish brown, but as with most Sea Urchins, individuals vary so in color that this is a poor recognition characteristic.

Farther south one encounters the Brown Urchin, *Cidaris tribuloides*, which ranges from the Carolinas to Brazil; its roe is highly appreciated in the West Indies, where this species abounds. The test is about 2 inches in diameter, and the spines are often 2 inches in length, making it a very long-spined urchin. I was amused at a description of its color found in the literature: "The color is light brown shaded with darker brown and sometimes mottled with white; some specimens vary toward olive-green while others are red. In some cases the spines are banded with purplish red and yellow." Just what color is a *Cidaris tribuloides?*

Again it is the Pacific coast that has the best Sea Urchin of all, the Giant Urchin, *Strongylocentrotus franciscanus*, and this urchin, being much larger than its small green cousin, is probably better able to bear that burdensome polysyllabic name. This is the favorite urchin of the Italians of California who first taught me to appreciate this luxurious food, and it is exceedingly fine. In late autumn and early winter, this species is often found with huge turgid roe sacs that are bright orange in color. The taste is very rich, without being in any way overpowering; in fact, it is quite subtle and delicate. One factor in this urchin's superiority is its great size, which means a worthwhile yield of roe from each individual. The test of this large species is often more than 5 inches in diameter. It has two color phases, red-brown and dark purple. The range is from Alaska to Mexico, though it is not plentiful in the southern part of that range. It is usually found in the deeper tide pools or just below low-water line on rocky coasts.

Over this same range is found the Pacific Purple Urchin, *Strongylocentrotus purpuratus*, looking very much like the above species, except for its size, for it is only about 2 inches in diameter. This little urchin actually manages to make excavations in the solid rock to give itself a secure resting place in the beating waves. The manner in which these relatively soft creatures manage to scoop out depressions in a medium much harder than any part of themselves is still hotly debated among naturalists. None of the theories advanced sound very convincing to me. A secretion strong enough to dissolve solid rock should be detectable in a laboratory, and the theory that they wear and bite away the hard rock with their comparatively soft teeth and spines doesn't seem likely. Some of these urchins imprison themselves in the rock by boring in when they are small, then enlarging the hole lower down as they grow larger. This is easily the most abundant species on the California coast. While it is much smaller than the big kind described above, it can be collected in such quantities that in most areas one could more quickly fill a basket with these small ones than with the rarer large ones. The roe, while small, is as delicious as that from the Giant Urchin.

Since Sea Urchins never try to run away or fight back, gathering them where they are abundant is about as much sport as pulling fruit from a tree. The best season to gather urchins on both the East Coast and the West Coast seems to be from July through December. At this time of the year a larger percentage of the urchins opened have full, orange-colored egg sacs, although even during this season, some will have limp and spent roes that are hardly worth removing. The yield of edible roe, when compared to the bulk of the live animal, is small. On one occasion I opened a 3-gallon pail of Green Urchins and procured only one cupful of roe. However, as would also be true of caviar, a cupful of urchin roe is a gratifying quantity.

To extract the roe, lay the urchin on its back, that is with Aristotle's lantern up, and crack the test all around with several sharp blows with a hammer. The lower part of the test and the viscera can then be removed, showing the 5-pointed egg sac, or gonads, against the upper part of the test. Reach in and loosen it from the points inward and it can be lifted out whole.

The best way to cook urchin roe is ... not to. The immigrants in California give it no preparation whatever, but merely eat it on crusty Italian bread. Some people, however, are squeamish about eating urchin gonads membranes and all. By experiment I found that I could put this roe through a fine sieve and remove the membranes; this leaves me with an orange liquid something like regular egg yolk. I tried whipping this with a power egg beater and it gradually became thick and spreadable and turned a lighter, lemon-yellow color. Eaten on small round crackers, or bits of toasted French bread, this Urchin Butter will make caviar taste like very common fodder.

Urchin gonads are eaten not only for their delicious flavor and healthful goodness, but also because of their great reputation as an aphrodisiac. I suspect that if this quality were subjected to a scientific investigation one would find that urchin roe has one thing in common with the rest of the dietary aphrodisiacs of folklore, and this is that it just doesn't work, or at most no better than any other rich and wholesome food. Its joys are those of the dining room.

14. The Purple Snail or Dog Whelk
(*Thais*, several species)

THE Dog Whelk, also called Rock Purple and Purple Snail, is one of the most common creatures of the intertidal zone on both the East and West coasts. *Thais lapillus*, the familiar Eastern species, crowds rocky shores from Long Island to the polar regions. It averages about 1½ inches in length and 1 inch in diameter and usually has a striated, thick-walled shell, a short spire, and a large, thick-lipped aperture, but they are extremely variable in size and shape, and especially in color. They are often found crowded together in great congregations on the shady side of rocks, in crevices, and in tide pools, and one will find in a single such colony individuals that are porcelain-white, lemon-yellow, orange, or purple-brown, and many are banded with two or more colors.

They are voracious carnivores, preying on barnacles, mussels, and even the eggs and young of their own species. They probably play an important role in keeping littoral populations under control. To get at its food, this resourceful and hard-working snail simply sits down on some shelled mollusk or barnacle and starts rasping with its file-like tongue, or *radula*, until it literally licks a hole through its victim's armor. It then inserts this sharp, hard tongue, slices up the soft, succulent creature within, and lifts the bite-size pieces into its own mouth. One often sees them in the process of penetrating a barnacle or mussel shell. The only trouble is that they seem to have no instinct for determining whether the shell contains a live animal or not, for they sometimes rasp holes in dead, empty shells; I have even seen one on the inside of half a dead mussel's shell rasping a hole toward the outside. What a disappointment it must be to complete so monumental a task and find nothing but water or beach sand on the other side! Maybe they are not hungry and are doing this seemingly useless work just to keep in practice.

The West Coast is well-populated with several species of *Thais*. The largest of these, *T. lamellosa*, can become 3 inches long and can acquire 9 to 20 axial processes on its shell, thin plates that are elevated and wrinkled at the ends where they overlap like shingles over a large spiral ridge. As would be expected of a creature carrying such fragile decoration, this snail is usually found in sheltered bays, from Alaska to Santa Barbara.

Thais lima is next in size, becoming over 2 inches long. It has a fairly uniform sculpture of alternate major and minor spiral ridges. It is found from Alaska to Baja California.

The Channeled Purple *Thais canaliculata* has a very uniform set of spiral ridges separated by distinct interspaces, without minor spirals. It ranges from Alaska to Monterey. Both it and *Thais emarginata* are about the size of the Eastern Dog Whelk, reaching a length of about 1½ inches. *Emarginata* is commonly called the Short-Spired Purple and has a fat, roughened shell; the color varies as does that of its Eastern relative, but is not so bright, being limited to whites, drab grays, and browns ranging almost to black. It is found from Alaska to Mexico. All these West Coast species are very similar to the Eastern Dog Whelk in their feeding and reproductive habits and in their possible uses to mankind.

Dog Whelks lay great numbers of eggs in little vase-shaped capsules with short stems and flat circular tops. These capsules are about ¼ inch high and are found in clusters on rocks, in crevices, and even on dead shells. Upwards of 50 eggs are deposited in each little closed vase, but they do not contain enough yolk to nourish the developing embryos until they are ready to break through the tough membrane of their nursery, so the first one hatched proceeds to eat the rest of the eggs and developing young, and then emerges into a world where it is still far from secure. A Dog Whelk could be the sole survivor

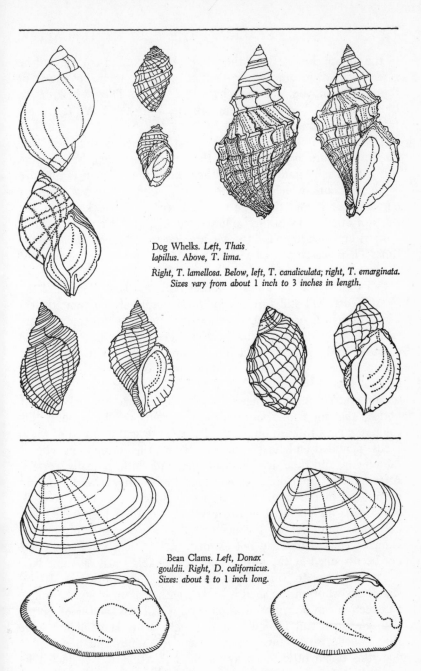

Dog Whelks. *Left, Thais lapillus. Above, T. lima.*

Right, T. lamellosa. Below, left, T. canaliculata; right, T. emarginata. Sizes vary from about 1 inch to 3 inches in length.

Bean Clams. *Left, Donax gouldii. Right, D. californicus. Sizes: about ¾ to 1 inch long.*

of the internecine battle of the natal capsule only to emerge and be eaten by its own cannibalistic mother or father. If, in the course of transmigration, you ever come back to earth as a Dog Whelk, try to be the first hatched out in your capsule, and then don't try to get friendly with your parents.

The common names Rock Purple and Purple Snail were bestowed on this creature, not because of the color of its shell, but because this is the snail family that furnished the famous Tyrian purple dye of the ancients, which sold for a king's ransom and was worn by emperors. It is said that one can still see, in the rocks about the ruins of ancient Tyre, the hollowed-out dye pits that were used as mortars in which to pound the Mediterranean species of *Thais*. Sometimes *Thais* was combined with certain *Murex* species to give different shades of this costly color.

Our own species of *Thais*, those of both the East Coast and the West Coast, will still furnish this royal purple, if you want to feel like a Roman Caesar. The secretion that provides the color is contained in a small gland. Strangely enough, this fluid is completely colorless, and the purple shade only develops when it is exposed to strong sunlight. I first noticed this purple dye while fishing on a sunny wharf. I was using Dog Whelks for bait, and after opening several and getting my fingers smeared with their internal fluids, I noticed that my fingers and nails were turning purple. I found this color very hard to wash off. Later, I opened several hundred of these shells, pounded and mixed the glands and the surrounding viscera thoroughly, and then dyed a small square of cloth. It really worked. A few minutes after the cloth was spread in the sun it began turning purple, and by the time it was thoroughly dried it was deeply colored. Still, I was disappointed in this royal purple, for I found the color dull and uninteresting. I concluded that the ancients must have known secrets that I had not learned, or else their color sense had not been spoiled by the brilliant dyes obtainable today, as mine has.

The chief use to which I put Dog Whelks is as fish bait. If the large body whorl of the shell is broken with a sharp blow, the tough foot section can be removed whole, and it will stay on a hook better than any good bait I know. I have caught many kinds of fish with Dog Whelks, so I know this bait has wide appeal. So persistently does this tough little foot stay on the hook that I once pulled in 23 smelts before I had to change baits. Another time, I caught 8 large

Pollack on one whelk within half an hour, and that is a good day's work for one small snail.

When I tour the shores of New England, I always carry a small bait box full of live Dog Whelks, so I am ready to stop and fish any likely-looking spot I pass. When I find a suitable rocky shore, I release the whelks I caught the day before and gather a new supply. The Dog Whelk can easily live out of water for several days, for it can breathe atmospheric oxygen that dissolves in the water retained about its gills.

As a source of human food, the various *Thais* species have definite possibilities. Though one should not expect this tough little snail to be among the finest of seafood delicacies, as plain, unexciting fare it is good enough to merit occasional use. When I broke all those shells to obtain the purple dye, I dropped the meaty foot sections into a bowl and saved them, for I was determined to test the utility of this marine snail in as many ways as possible. The little disks of meat average about ⅝ inch across and ⅜ inch thick, so even with several hundred of them I had no trouble finding a container large enough to hold them. Attached to the foot is a horny, brown operculum, which the snail uses to close its door when it wants to go inside and retire. This thin plate is easily lifted off with the blade of a knife, or even with the thumbnail.

I washed these meaty sections thoroughly, then bravely popped one into my mouth raw. It didn't taste at all bad, but I was still enjoying it half an hour later. Boiling a few didn't make them much tenderer, but they were no worse in this respect than unpounded Abalone. I put the remainder of the meat through a food chopper as I have recommended for other tough mollusks, and this seemed to be the answer. I mixed some of the ground Dog Whelk with a little beaten egg, then made it into little patties and fried them. It was distinctly good. Next I mixed the ground meat with finely chopped onion, celery, green pepper, and bread crumbs in the proportion recommended for Stuffed Quahogs, page 21. This mixture was mounded on the large scallop shells that I use for baking, then each mound was topped with a slice of bacon, and they were baked at 450° until the bacon was crisply done. These were actually delicious.

On another occasion I gathered two quarts of Dog Whelks and used them in a chowder. These were steamed in a covered kettle for about 15 minutes, when most of the opercula dropped off and the ones left were easily removed. The meaty foot sections were fished out of the shell with a nutpick, and the shells and viscera were discarded. The two quarts of snails yielded only about 1 cup of edible meat, which I put through a food chopper and then made into a chowder, using the recipe for Quahog Chowder that I give on page 21. While not in the same class as the chowders made from Coquinas, Bean Clams, or Surf Clam muscles, it was still a good chowder.

15. Coquinas and Bean Clams:

THE WEDGE SHELLS

(*Donax*, 4 species)

"HOW small does a clam have to be before you will let it alone?" This question was put to me by a friend who was watching me collect tiny Bean Clams, *Donax gouldii*, from the sands of Long Beach, California. I must admit that at first glance this clam seems altogether too small to be considered a food species. And yet, these miniature mollusks were formerly eaten and highly appreciated, and at one time formed the basis of a canning industry that employed a large number of people. These are solid accomplishments for a clam so small. Incidentally, that evening, after my friend had eaten a bowl of delicious Bean Clam Chowder he withdrew his facetious question and promised never again to ridicule the Bean Clam because of its small size.

My friend had asked the wrong question. The size of the individual units of any food product is not significant. A grain of wheat is pretty small, but its products are the staff of life over much of the earth. The pertinent questions when enquiring about an edible mollusk are: "What is its quality?" "What is the proportion of edible food to shell?" and "How many pounds can be gathered in an hour?" In these respects the Bean Clam comes through with flying colors, and this little clam really has some colors to fly.

The average Bean Clam is only about ¾ inch long, and its form is strikingly different from that of other clams we have discussed. The valves are heavy and strong for their size, deeply arched, and marked by indistinct radiating lines and concentric growth lines. This clam is distinctly wedge-shaped, that is, the anterior or foot end is elongated to a thin edge, and the posterior or siphonal end is abruptly truncated, or flattened, appearing cut off just aft of the hinge and forming a right angle with the hinge line. The outer margins of the valves are heavily crenulated, and these protuberances mesh or interlock when the shell is closed. The color is variable, from white to blue to purple. Some specimens are conspicuously marked with concentric bands of color, and others have rays of color extending from the hinge section to the shell margin. Occasionally one finds an individual with both rays and concentric bands forming a plaid design in three colors.

This colorful little clam is found in smooth-packed sand on the outer coast from Santa Barbara, California, to Acapulco, Mexico. While it can survive on surf-pounded shores, it seems to prefer some slight degree of protection, and congregates most thickly about the mouths of bays and behind offshore sandbars. They are found at the surface, or barely buried, from mid-tide level to below the lowest tides. Their siphons are separate and very short; consequently, they cannot burrow deeply and survive. Often they will be found on hard-packed sand with the blunt siphonal end projecting slightly above the sand. On many specimens the blunt end is provided with threadlike wisps that clam diggers call "whiskers." These whiskers are really separate groups of animals, being the plantlike colonial hydroid, *Clytia bakeri*, which also attaches itself to the shells of the Pismo Clam. It seems to be the only hydroid that can survive on surf-pounded, sandy shores. It attaches itself to the smooth shell of the Bean Clam just as it would to a rock or any other suitable surface. Often one sees little tufts or patches of "clam's whiskers" protruding from the smooth sand, thus betraying the presence of many Bean Clams just under the surface.

Formerly these clams were exceedingly abundant on the hard-packed sand beaches of southern California. In 1895 a cannery was set up to produce what was called "Clam Extract" from this species.

A great many boys were hired to collect the clams. They pushed two-wheeled carts with mesh bottoms to the beaches and filled them with the clam-crowded top layer of sand. The carts were then covered with a screen and pushed into the surf, which soon washed the clams free of sand. Using this method one could collect in a few minutes more clams than could be carried away. After even a small storm, live Bean Clams were heaped up in long windrows. At these times the boys could shovel almost pure clams into their carts, and they reaped a rich harvest. Thousands of pounds were daily taken from the beaches and processed in the cannery. A contemporary account says, "These shellfish are so plentiful that the same ground is worked over day after day; if the beach is gleaned one tide, the next leaves a fresh supply."

I wonder where those people thought this seemingly inexhaustible supply of clams was coming from. In fact, I also wonder where they *were* coming from, for despite this intensive exploitation, the Bean Clam boom lasted an amazing fifteen years. Then, either something happened to the clams, or else the mighty supply was finally exhausted, for the clams suddenly disappeared and the cannery had to shut down for want of raw material. Since then, they have been more abundant some years than others, but they have never regained their former numbers, and today almost no use is made of them. Even where they are available, practically no one collects them, for Southern Californians have long since forgotten that they once considered this little mollusk the most delicious of all clams.

It is no longer possible to gather bushels of Bean Clams in a few minutes, but the alert forager can still find enough to make many a good soup, chowder, or nectar. There are many areas in Southern California and Lower California where these clams can still be found in worthwhile numbers. Each shovelful of surface sand may now contain a dozen or fewer, where once it would have contained hundreds, but a clever clam digger can still get a gallon or two of Bean Clams during one low tide, if he knows where to look. Enough for a good pot of soup can sometimes be gleaned by combing the surface sand with a fine-toothed rake or even with the fingers, but the old-fashioned method of washing them from the sand is still the best way to get them in quantity. I've never tried to revive the two-wheeled washing-cart from days of yore, but used the same galvanized, hardware-cloth basket that I used in de-mudding Bent-Nose Clams (described on page 179). I first look for tufts of revealing hydroids, or "clam's whiskers," and on finding goodly clusters I fill the basket with surface sand from that spot. Then I wade into the water, slosh the basket up and down until the sand is washed out, then dump my catch of clams into a pail. Sometimes a basketful of sand will yield only a few dozen little clams, and other times there will be more than a pint. Each washing, large or small, adds to my hoard, and by the time the returning tide chases me from the beach, I have usually filled a two-gallon pail to the brim.

The Bean Clam is so small that opening it while raw would be too tedious a task, so this species is not suited for making fried clams, clam fritters, or for eating raw, but it does make superb soup. We will discuss cooking methods further after I have introduced some of the Bean Clam's near relatives, for they are much alike—all small, all equally delicious—and all can be cooked in the same manner.

From San Pedro southward, over the same range occupied by the Bean Clam, one also finds the closely related species, *Donax californicus*, which is also called Bean Clam because most clam diggers do not distinguish it from the species described above. It is even smaller than *gouldii*, seldom exceeding ½ inch in length. It too has a flattened posterior, but this plane is not so acutely angled in relation to its hinge line, so it doesn't have the suddenly-cut-off appearance of the abruptly truncated Bean. While *gouldii* has its hinge almost against its truncated posterior end, *californicus* has its hinge located near the center of the dorsal side, as do most clams. Otherwise it has much the same appearance as the first Bean Clam described, with the same heavy shell with crenulated margins, but its colors are usually more drab than those of its gay kinsmen. I have never really tried to collect *californicus* but I have sometimes found numbers of them in my catch of Bean Clams. They seem to prefer a more protected habitat than *gouldii*, and where spits or sandbars keep the surf from breaking on a sand beach, this species sometimes predominates. I never remove this species from a mixed catch, and since I have never noticed any debasing of the flavor of the soup, I imagine we can assume that it is just as delicious as its larger relative. It is only mentioned here so you will not worry if you find some of the clams in your catch do not conform to the description of Bean Clams.

When we turn to the East Coast of the United States, we find a strange parallel between the *Donax* species of our Atlantic shores and those of the Pacific. Here again there are two common species, one about ¾ inch long and brightly colored, and the other smaller and duller in hue. The larger of these two, *Donax variabilis*, is so similar to the Bean Clam of California that two names hardly seem necessary. There are technical differences that separate them into two distinct species, but they are minor. The main thing that separates them is a very large continent.

Such parallel distribution is not unusual among seashore animals that range into the Arctic. Such species as *Mya truncata*, a close relative of our common Soft Clam, and the little Green Sea Urchin with the tremendous name, *Strongylocentrotus drobachiensis*, and many other Northern species are found on both the Atlantic and the Pacific shores of North America. But these are also found in more or less continuous distribution around the top of this continent, through the Arctic Ocean and the Bering Sea, and around that great peninsula that is Alaska, so we can easily surmise how these creatures made the trip. Obviously many life-forms discovered and used the famous Northwest Passage ages before mankind even started hunting for it. Although this is very interesting, it does nothing to explain the two-ocean distribution of *Donax*, for this genus has no Arctic species.

The fact that Eastern and Western species of *Donax* are so closely related is only one item in a vast body of evidence that has convinced scientists, almost to the point of certainty, that the Atlantic and the Pacific were connected between North and South America at not too remote a time. The creatures apparently looked much as they do today when they passed from ocean to ocean through this Pliocene Panama Canal.

Donax variabilis, the Eastern equivalent of the Western Bean Clam, has a number of common names, including Coquina, Pompano, Butterfly Shell, Calico Clam, and Variable Wedge. It is found from Cape Hatteras to Florida and around the Gulf Coast to Texas, and it is one of the handsomest and daintiest of bivalve shells to be found in this area. There are some differences between it and the Western Bean Clam that even an amateur can see. The flat part of the truncated posterior end is set at a more oblique angle to the hinge line than it is on the almost right-angled Bean. The coloring is like that of the Bean Clam, only a great deal more so. The specific name *variabilis* pertains to its variable coloring, and in a random handful of Coquinas you are not apt to find any two colored alike. Colored rays diverge from the hinge section to the shell margin to be crossed by concentric bands of harmonizing or contrasting hues, involving all the colors of the rainbow and some the rainbow never thought of wearing. There are whites, pinks, rose, blue, lavender, green, yellow, fawn, purple, red, and brown worked up into patterns of plaids, stripes, calicos, checks, and some that defy description. Even the glossy inside of the shell is usually suffused with color. Biology teachers often mount as many as fifty different-colored Coquinas on a board to demonstrate variability within a single species. Naturally, curio stores and gift shops work these pretty little shells into all sorts of incongruous and so-called ornamental objects, but these I can take or leave alone. I much prefer the simple, nature-embellished shell that I find alive on the beach.

I don't think the Coquina has ever been sold commercially as a food clam, but residents of areas where it abounds long ago discovered that this tiny clam makes excellent broth, soup, or chowder. There are places in Florida where one can pick up a handful of the beach and have more clams in one's hand than sand. Using a box with a bottom of coarse screening, it is possible to wash out many quarts of these clams in a few minutes. I once watched a woman get enough for a huge pot of Coquina soup in a very short while, using an ordinary flour sifter to separate them from the sand. When you are searching for seaside delicacies in the South, don't overlook the little Coquinas.

The only other common species of Wedge Shell found on the East Coast is *Donax fossor*, commonly called the Digger Wedge, found from Long Island to Florida. This is a small member of a family of very small clams, averaging only about ½ inch long. Most people would think that its small size would make it safe, even from me, but not so. It too has a flattened posterior, but this flattened part of the shell is rounded into the other parts instead of being acutely angled as in other Wedge Shells. It would be considered a colorful little creature if it didn't have such gaudy relatives. The most common color is olive decorated with bluish rays. The pattern varies from one individual to another, but the Digger has never achieved anything like the variability of its cousin, Coquina.

This tiny Digger Wedge is easily the most abundant bivalve on the New Jersey shore if one considers only numbers of individuals.

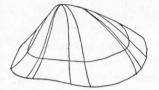

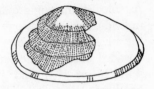

Left, Coquina. Length: about ¾ inch. Right, Digger Wedge. Length: about ½ inch.

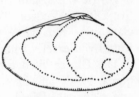

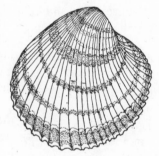

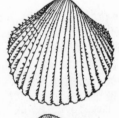

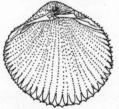

Basket Cockle. Exterior and interior views of shells. About 3 inches in diameter.

Spiny Cockle. Exterior and interior views of shells. About 2 inches in diameter.

Rock Cockle
Exterior and interior views of shells. About 1½ inches in diameter.

A biologist, making a quantitative study of these clams, counted the individuals found on one square foot of the lower beach and found the amazing number of 1,510 living Diggers in this small area. Once I waded out to where some newly-driven piling protruded from the water to see what kinds of sea creatures were beginning to colonize on them. Where the piling entered the sand the tide had washed out little pools, and in each of these pools were a quart or so of live Digger Wedges. That evening I enjoyed a wonderfully savory clam soup. This happened in Cape May County, but I'm not about to tell you exactly where.

The Digger Wedge has a much thinner shell than does the Coquina, so, despite its small size, it has a larger proportion of flesh, and a quart of Diggers will make a stronger broth than will a quart of Coquinas. The simplest way to cook this clam, or any other species of *Donax*, is to add 2 cups of water to 2 quarts of washed clams and steam gently for 20 minutes. Strain off the broth and throw the clams away. There is only an infinitesimally small bit of flesh in each tiny shell, and even that is likely to be gritty. The broth can be served in small glasses or demitasse cups, and called Clam Nectar, or it can be served as a clear soup at the beginning of a seafood meal.

For a little fancier product, add ½ cup dry white wine and 1 cup water to 2 quarts of clams in the shell, and steam for about 15 minutes. This Wine Nectar can be served like the broth above, or it can be used as the base for a delicious soup. In making Coquina Soup, I heat 1 cup of milk and 1 cup of light cream in the top of a double boiler. While this is heating, I cream 2 tablespoons of flour with 2 tablespoons softened butter until I have a smooth paste. I add this to the winy broth and cook and stir until it thickens slightly. Then I add the heated milk and cream, and season with a dash of freshly ground black pepper and a tiny dash of nutmeg. Just before serving, I stir in 2 tablespoons cooking sherry. One of my friends who is a chef says that I should never mix wines in this fashion, but I find it perfectly delicious and there is no law against it that I can discover.

I have made chowder in many ways from Bean Clams, Coquinas, and Digger Wedges, and all of them were very good. It was in an attempt to develop a chowder that was something more than merely "very good" that I worked out the recipe given below. This is the sort of thing you do when the creative urge is on you and you want to elevate a chowder from an ordinary soup into the realm of liquid confections. I call it Ne Plus Ultra Chowder, and I made it of Digger Wedges, but it would be as good made of any species of *Donax*.

Add ½ cup of dry white wine and 1 cup of water to 2 quarts *Donax* clams of any species, simmer for 15 minutes to extract the broth, then strain, discarding the clams and shells. Fry 4 slices of bacon until crisp, drain on a paper towel, and when cool crumble finely. You will need 1 cup diced boiled potato and ½ cup chopped leeks, using only the white part. If you can't get leeks, use the white part of green onions, or even chopped white onions. Cook the diced potato and chopped leeks in the bacon fat until the potato is a delicate brown and the leeks are clear and yellow. Drain on a paper towel to remove as much fat as possible. Take 6 ears of freshly gathered sweet corn, slit each row of grains down the center, then with the back of the knife scrape down the rows of grains, pressing the insides out into a mixing bowl. Add the corn cream, bacon, potato and leeks to the broth, reserving ½ cup of the broth to mix the thickening. Bring the broth-vegetable mixture to a boil, then reduce the heat and barely simmer for 20 minutes. Heat 1 cup of light cream and 1 cup of milk in the top of a double boiler. Using a small amount of the reserved broth, stir 4 tablespoons flour into a smooth paste, then mix this paste with the remainder of the reserved broth. Stir this into the pot of soup, and keep stirring until it thickens slightly. Add the hot milk and cream, and stir again. At this point taste it to see whether or not the clams have contributed enough salt to make it taste right. If not, add salt to taste and ⅛ teaspoon freshly ground black pepper. Bring just to a simmer and serve, sprinkling each serving with finely chopped parsley and chives. With this chowder serve little round pilot biscuit, Beach-Plum Jelly, and a small glass of the same dry wine you used in extracting the broth. Delicious.

16. The Cockles of America

THE English language is deficient in common names to apply to edible mollusks. There are clams, mussels, oysters, scallops, and a few others, but into this paltry handful of terms we must crowd a vast number of often unrelated life-forms. To distinguish between the different kinds, we add a descriptive word or phrase; for instance we have Soft Clams and Hard Clams, Littleneck Clams and Big-Neck Clams, Horse Clams and Otter Clams, Razor Clams and Jack-knife Clams, Bent-Nose Clams and Round Clams, Mud Clams and Sea Clams, to name only a few. This would not be too bad if we remained consistent, but we don't. The Hard Clam of one part of the country might be called the Soft Clam in another area where it is being compared with a different species. Some common names are so local in use as to have little significance, while another name, widely used, may be applied to a different creature in each locality. For these reasons we must refer to those hard-to-spell Latin scientific names to be sure that you and I are talking about the same beastie when we are discussing any edible mollusk.

A good illustration of our overuse of even a subordinate term is the way we use the common name "cockle." This was originally the common name of only one species of *Cardium*, or heart shell, found along the shores of the British Isles. The term became widely known because cockles were eagerly sought, widely marketed, and greatly esteemed at the table. These were the cockles that were sold in "Dublin's fair city" to the cry of "Mussels and cockles, alive, alive-O!"

When the English-speaking people spread all over the earth, the term "cockle" was broadened to include all species of *Cardium* wherever found. Later it began to be applied to almost any clam with a real or fancied resemblance to the *Cardiums*. Today on our own West Coast there are dozens of species of four different genera that are commonly known as "cockles." A term so broadened ceases to tell us very much about the creature it refers to, so we are forced to add modifying words to the overcrowded term, thus creating our own binomial system of common names. In due course the people along the Pacific coast began to speak of Giant Cockles, Spiny Cockles, Basket Cockles, Sea Cockles, Rock Cockles, Hard Cockles, and Soft Cockles.

You will note that at one place above I referred to the cockle as a clam. The term "clam" also originally referred to a single edible species found in Britain, but it has since been broadened until now it gives much greater coverage than even the term "cockle." In fact, the popular usage of the term "clam" includes the cockles, and cockles are really only one family of clams or bivalves. This illogical usage of common terms may puzzle the casual beachcomber a bit, but those who make a deeper study of this subject are soon lost in utter confusion.

Along the Atlantic seaboard the term "cockle" has largely been reserved for the species of *Cardium*, the genus that contains the original cockle and therefore probably has the best claim to the name. There are some cockles along our northeastern coasts, but either they are too small to interest us or they live in water too deep for us to get at them. However, along the shores of our southeastern states there lives one of the largest, finest cockles in the world, *Dinocardium robustum*. This beautiful big cockle, known as the Giant Atlantic Cockle, is found from Cape Hatteras to Mexico, and on some of the Florida beaches it becomes abundant. Large specimens measure more than 5 inches from hinge to lip and more than 4 inches across the other way. The *"cardium"* part of its name means heart-shaped, and this cockle viewed end-on presents an outline that is an almost perfect St. Valentine's heart. Looking at the side, one sees that the posterior or siphon end is abruptly truncated or flattened, while the anterior or foot end is gracefully rounded. The shell is sculptured with from 33 to 37 (usually 35) broad, regularly disposed, radiating ribs that create a beautifully crenulated margin where they end at the lips of the shell. The shell is yellowish brown in color, with transverse lines or rows of purplish-brown spots, and the flattened posterior end is purplish brown. The foot is large, curved, and pointed at the outer end, and the siphons are so short as to be mere openings in the edge of the mantle. These short siphons keep the cockle from burrowing deeply, and at extremely low tides they often lie partly exposed where they can be picked up without digging.

The flavor of the Giant Atlantic Cockle is good, but a bit strong. I care little for them steamed or fried, and they are very poor in a Clambake, but they make good chowder and are superb as Stuffed or Deviled Clams. The large, deep, and beautiful shell immediately suggests itself as the proper baking and serving dish in which to present these clams. Half a dozen cockles will make enough Deviled Cockles to serve two people lavishly. Open cockles by inserting a thin knife between the valves and severing the hinge muscles. Remove the meat and grind it coarsely. Add 1 beaten egg, 2 tablespoons tomato catsup, 1 teaspoon horseradish, 1 teaspoon minced parsley, 1 tablespoon minced onion, 1 tablespoon minced celery, ¼ teaspoon ground thyme, ⅛ teaspoon freshly ground black pepper, and 1 cup bread crumbs. Mix all ingredients together, and stuff 2 of the largest cockleshells. Cover each mound with half a slice of bacon, and bake in a medium oven until the bacon is just crisply done. If you dislike this dish, your taste and mine are too divergent for me to advise you on food.

There are several other species of cockle on our southeastern shores, especially in Florida and around the Gulf Coast. I'm sure that all of them are edible, whether they are commonly eaten by the local populace or not. The only one that I have experimented with besides the Giant Atlantic Cockle is *Trachycardium egmontianum*, commonly called the Prickly Cockle, or sometimes the Red Cockle. It is proportionately much higher from lip to hinge than most cockles, an average specimen being about 3 inches high and only about 2 inches across. The shell is brownish to straw-colored,

decorated with purplish spots; there are from 27 to 30 prominent
ribs, and on the tops of the ribs are small, erect scales marking
the concentric growth lines. When opened, this cockle is unmistakable, as the inside of the shell is salmon or purplish pink. Like
the Giant Atlantic Cockle, it is not particularly good steamed or
fried, but it makes number-one chowder or Stuffed Clams, or you
can use the recipe below.

Clam Potato Croquettes are a good way to use cockles or tough
clams. Open a dozen Prickly Cockles and save the liquor. Melt 1
tablespoon of butter in a saucepan and stir in 2 tablespoons of flour.
Cook and stir until the flour starts to brown, then pour in the clam
liquor and about ½ cup of milk. Cook and stir until the sauce
thickens, then cool. Grind the meat from the cockles with 1 cup
cold boiled potato, chop 2 slices bacon and 1 onion, and fry until
the onion turns yellow and the bacon is nearly done. Mix all ingredients together, add 1 beaten egg and about ½ cup bread crumbs.
Allow to cool, then shape into croquettes, dip in egg beaten with a
little water, and roll in bread crumbs. Fry in 375° fat until each
croquette is barely a golden brown. This good hearty chow will be
appreciated after a hard day of combing the beaches.

As with many other seafoods, the West Coast forager has the
advantage of the Easterner in the matter of cockles. The *Cardiums*,
or "true" cockles, are well represented along Pacific shores, but
where they are scarce the Westerners have merely rechristened some
other creature and gone ahead enjoying "cockles" anyway.

The most important, though not the largest, of the true cockles of
the West Coast is the Basket Cockle, *Clinocardium nuttallii*, found
from the Bering Sea to Lower California, but reaching its maximum
development in the Puget Sound–British Columbia area. Though
not a giant like *robustum*, this is a good-sized cockle, averaging about
3 inches across, and it has the typical heart shape when viewed
end-on. The grayish-brown shell is decorated with strong ribs and
decided valleys between them, all about equally spaced. Where
these radiating ribs and valleys join the margin they form interlock-
ing points with the other valve, as in the scallop.

The foot is very large, pointed, and with an abrupt elbow-bend
about the middle of its length. Siphon tubes are completely lacking,
there being only 2 openings in the margin of the mantle to replace
them. It is only clams with long siphon tubes that can burrow

deeply and still reach back to the surface for the oxygen-and-food-bearing water that is vital to their existence, so the cockle, with no siphon tubes, must remain very near the surface.

Basket Cockles are usually found on the shores of protected bays, often in coarse sand that would be very shifty if the surf could get at it. You need not dig these cockles, as they lie about partly or wholly exposed, and in some areas you can, with very little effort, pick up as many as any reasonable person would want during any low tide. One pleasant feature of this kind of cockling is that all the Basket Cockles you find will be of edible size, without the preponderance of undersized ones encountered when seeking other kinds of clams. A scientist of British Columbia who has studied the natural history of this species explains how this is possible. The Basket Cockle is a true hermaphrodite, shedding both ova and sperm in astronomical quantities during a long breeding season each spring. The fertilized ova hatch into free-swimming larvae and these head offshore and settle to the bottom below the lowest tides. The cockle is not a fixed bivalve but moves about freely, making hops of several inches at a time by suddenly thrusting its strong foot down and backwards. By this one-legged mode of progression, the young cockles gradually work their way toward shore, losing many of their number to hungry fish, crabs, and other predators they encounter on the way. Finally the few survivors enter the intertidal zone when they are two to three years old and just the right size to make excellent food for human beings.

A friend of mine, after reading the above paragraph, began bemoaning the hard lot of the Basket Cockle. He saw the migration of the young cockles from the deeps to the shore as a tragic Hegira, their numbers being constantly reduced by the finny denizens of the deep and the shelly monsters of the shallows. Finally a pitiful remnant of the once great horde hops into what should be the comparative safety of the intertidal zone only to meet those bipedal monsters, men, most voracious predators of all.

All this is sheer anthropomorphic nonsense. Man cannot and must not interfere with what my friend considers the appalling death rate among young cockles. If only ten out of each million cockles that are hatched managed to survive that hazardous migration to the shore, the beaches of the whole world would soon be mountains of cockles. Let us rather think of these fascinating creatures as forming vital links in some of nature's marvelous food chains. Even the microscopic larvae furnish essential food to plankton-sized crustaceans which in turn furnish food for small fish which in turn furnish food for larger fish which in turn ... Before the shells of the survivors grow too strong, they furnish food for crabs and other larger crustaceans, and the adolescent cockles are a favorite food of several valuable kinds of fish. The remnant of the original vast number of cockle young finally hops into the intertidal zone and spawns, to assure the continuance of its own race and the continuance of an important food supply for many other species. Here we may enter the picture, and some of the adult cockles go to make a delicious bowl of chowder for you and me. There is nothing tragic about a creature playing the role that nature has assigned to it, and the cockle plays at least its final role to glorious perfection.

Basket Cockles have a good clammy flavor but, like the Eastern Quahogs, they are inclined to be a bit tough. For this reason they are better used in dishes where they can be ground or chopped before cooking. I seldom go hunting solely for Basket Cockles. When I am digging Butter Clams, I always find a few Basket Cockles lying about. I welcome them into my clam basket, for these two species blend well; chowder of mixed Butter Clams and Basket Cockles is better than one made of either species alone. Grind the meat from the cockles, but leave the Butter Clams whole when making this superior chowder. Otherwise follow the recipe on page 21. The two also harmonize beautifully in Stuffed or Deviled Clams, and you will surely want to use the large, beautiful cockleshells as baking dishes rather than the small plain shells of the Butter Clams.

If you wish to try a dish that uses Basket Cockles without the addition of any other kind of clam, Cockle Bisque is guaranteed to please. Open a dozen Basket Cockles over a mixing bowl so you can catch the liquor. Add enough water to the liquor to make 1 cup of liquid. Grind or chop the meat. Add the cockle meat, ¼ cup diced celery, and 1 tablespoon chopped parsley to the liquid, bring to a boil, then turn the heat down and simmer for ten minutes. Press through a sieve and keep hot. In another saucepan melt 1 tablespoon butter, add 2 tablespoons flour and ⅛ teaspoon freshly ground black pepper. Quickly stir into a smooth paste, and cook and stir for 2 minutes, then add 1 cup light cream and cook and stir until it just comes to a boil. Quickly add the strained clam mixture and as soon as it again comes to a simmer, serve in individual ramekins with a round of buttered toast floating on each serving. This recipe will furnish enough for a light soup course for four or a real meal for two.

I have been able to gather enough cockles for a good chowder as far south as Monterey Bay, but south of there they become scattered and rare. From Santa Barbara southward to the southern tip of Lower California, one sometimes finds specimens of two perfectly enormous species of cockle, but unfortunately both of them usually stay on bottoms too deep to be available to the average forager. However, the growing popularity of scuba-diving gear makes these undersea clam beds more available to us than they once were, and a diver who brings up either of these oversize cockles should know that it is good to eat. The Spiny or Giant Pacific Cockle, *Trachycardium quadragenarium*, is about the size of the large Eastern *robustum*, but lacks the flattened posterior end of that species and has a plain yellowish-brown shell. This shell has the characteristic heavy radiating ribs of the typical cockle, but differs in having these ribs decorated with small spines or "thorns" at the growth lines. The Pacific Egg Cockle, *Laevicardium elatum*, is even larger, sometimes being as much as 7 inches across. The large size of these two cockles will serve to identify them, while the presence or absence of spines will enable you to distinguish between the two species; the Atlantic Spiny Cockle is much smaller. Both are eminently edible, and those large, capacious shells fairly cry aloud to serve as decorative baking dishes in which to prepare some stuffed cockles.

Over much the same range where one finds these two huge cockles, one sometimes finds another large bivalve, *Amiantis callosa*, called the Sea Cockle, though how this clam ever earned the title "cockle" I'll never know. It is not a *Cardium* and does not even have the radiating ribs one expects to see on a creature called "cockle." It is 4 to 5 inches across, with a dirty white, very heavy shell similar to that of the Pismo Clam. The shell is sculptured with concentric growth lines, but these are not evenly spaced and sometimes seem to fork or run into one another. It has a regular siphon instead of the mere mantle opening of the *Cardiums*. It is found on open coasts or at the mouths of bays, and doesn't hide in water quite as deep as do the two species described above, being found at or just below low-tide line. Look for them on open coasts and near the mouths

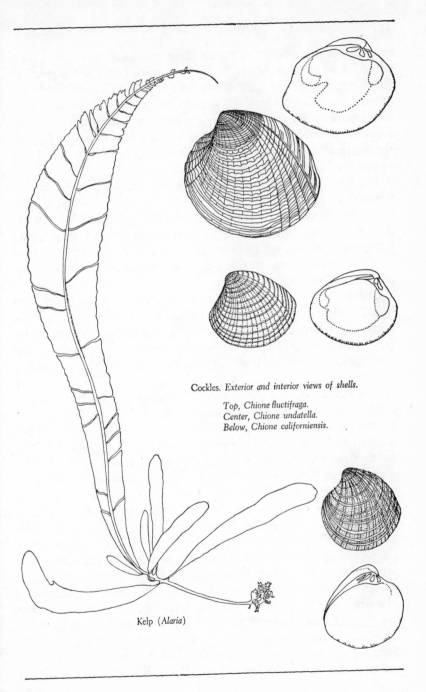

Cockles. *Exterior and interior views of shells.*

Top, Chione fluctifraga.
Center, Chione undatella.
Below, Chione californiensis.

Kelp (*Alaria*)

of bays, but don't expect to gather them by dozens anywhere. I have never found an area where these clams were really abundant, but this is not to say that such places do not exist in Lower California beyond the range where I have collected. Usually these fine, large clams come to me one or two at a time when I am really searching for some other species, but I welcome these few and add them to the pot and have always found them very good eating. Use them as you would use any cockle or large clam.

The Rock Cockle, *Protothaca staminea*, is a relative of the Eastern Quahog that is called a cockle because its shell is decorated with prominent radiating ribs like some of the *Cardiums*. It is found from Alaska to Lower California, but its distribution is very discontinuous, being abundant enough to assume commercial importance in several widely separated areas, yet being often totally absent on long stretches of coast in between.

The Rock Cockle is a clam of moderate size, 2 to 3 inches in length, but the valves are deeply arched so that the whole creature is rotund in form and of fair weight for its size. The shell is chalky, rather than polished, and is usually off-white in color, although occasionally one finds an individual with interesting zigzag markings in brown.

The Rock Cockle is ill adapted for rapid burrowing and could never maintain itself in the shifting sands that are the favorite habitats of the Western Razor Clam and the Pismo Clam. Rock Cockles are never found in pure sand or in soft mud. Look for them on firm gravelly beaches or in small accumulations of the claylike gravel found between the rocks on boulder beaches. In Southern California these clams are often found associated with the Hard Cockles described below, and in the north they are often found in the company of the Butter Clam.

Although Rock Cockles are found in commercial quantities mainly on protected bays and inlets, they are also able to maintain themselves on the outer coast wherever there are pockets of firm claylike gravel between the boulders along a rocky shore. Such small, isolated clam beds could never be profitably exploited by commercial clam diggers, but they can be of importance to the seaside camper or coast dweller. When walking along a rocky shore at low tide, keep your foraging eye peeled for low gravelly spots where an inch or so of water stands when the tide is out. I have discovered such potholes where Rock Cockles lay so thick just below the surface that their valves were touching one another, and a spade thrust down would bring up more clams than gravel. One such depression only a few square feet in area will often yield more Rock Cockles than a hungry family of campers can eat at one sitting.

The Rock Cockle is sweet and tender, and many West Coasters rate it very high in flavor. Probably the best way to eat Rock Cockles is as Clams Casino (page 21), for the flavor of this good clam seems to be completed and fulfilled by the addition of a bit of smoked-bacon taste. They are also excellent steamed or fried, and they are very good if you just roast them on a charcoal grill or in an oven until they have opened, then serve on the half shell with Melted-Butter Sauce. They can also be used in chowder or bisque, and one could convert them into Stuffed Clams or Deviled Clams. No matter how prepared, they are sure to please anyone who enjoys the flavor of good, fresh clams.

The clams that are known as Hard Cockles on the West Coast are really three species of *Chione*, and they are more closely related to the Eastern Quahog than to the true cockles or *Cardiums*. These three species occupy the same range and are often found together; all three have the same common name, and I have seen them mixed together in the Los Angeles markets. They are so nearly alike that they can be treated together here. In use and in flavor, they seem to be identical. All are comparatively small, from 1½ to 2½ inches in diameter, compact and rounded in outline, with firm heavy shells and short united siphons. The differences among the three species are chiefly in surface markings, as may be seen from the illustrations on page 157. There is really no reason why a seaside forager should attempt to distinguish between them, for all three are equally edible and, to my taste at least, their flavors are identical.

Chione fluctifraga reaches the largest size of any of the three and is probably the species that earned the name "cockle" for this genus, because it is distinctly heart-shaped when viewed end-on and has prominent radiating ribs like those of the *Cardiums*. In both *Chione undatella* and *Chione californiensis*, the radiating ribs are there, but they are overlaid by the more prominent concentric growth lines.

These three clams are distinctly Southern forms, being found from San Pedro to the tip of Lower California. They prefer protected sand flats that are not too much disturbed by surf or tidal currents.

They seem able to stand a pretty large admixture of mud in their habitat, and I have often found them on flats that would be better described as mud flats than as sand flats. Having short siphons, they do not burrow deeply and so are often seen with the siphonal end exposed.

After the Pismo Clam, these Hard Cockles are the favorite clams of Southern Californians; they make excellent chowder or bisque, and, unlike the *Cardiums*, they are also very good steamed or fried. They would make an excellent New England–style Clambake, and a favorite way of preparing them is as Savory Clams en Casserole. To make this last dish, steam 2 quarts of Hard Cockles in a covered kettle, using ½ cup white wine to start the steaming. Remove the clam meat from the shells and reserve the liquor. Dice ½ pound of fresh mushrooms, preferably wild ones. Pasture or meadow mushrooms work well in this dish, but even better are shaggy-manes, inkies, young oyster mushrooms, or puffballs. Of course you can substitute domestic mushrooms from the market, just as you can substitute almost any species of clam for the Hard Cockles. Such purchased ingredients will still taste good and will feed your body well, but unless you gather the main ingredient yourself, this dish will do very little for your soul.

Sauté the chopped mushrooms for just 2 minutes in a covered kettle. Drain the juice into the clam broth. Mix the chopped mushrooms and clams together, and season with 1 small onion chopped, ⅛ teaspoon ground thyme, and 1 teaspoon finely chopped parsley.

Make a sauce by melting 4 tablespoons butter in a saucepan, then stirring 4 tablespoons flour in it; keep stirring until it is a smooth brown paste. Add the clam juice, mushroom juice, and ½ cup milk, and cook, stirring frequently until it thickens. Stir in the seasoned clams and mushrooms.

You now have a versatile mixture that can be served in any of three ways. It is very good just served "as is" over buttered toast; or you can pour it into patty shells and brown for about 2 minutes under the broiler before serving; or you can pour the entire mixture into a buttered baking dish, cover it with buttered bread crumbs, and cook in a medium oven until the crumbs are browned. Even better, pour this savory clam mixture into some large cockleshells, cover it with buttered crumbs, and bake until it is nicely browned on top. Served in any of these ways, it is sure to be appreciated.

17. The Pen Shells, a Favorite Food of Ancient Gourmets

(*Pinna* and *Atrina* species)

THE Pen Shells are widely distributed and have been highly appreciated as food mollusks since ancient times. Indeed, to judge by the number of times this creature is mentioned in ancient writing, it must have been a favorite shellfish of Greek and Roman epicures, served at all the finest banquets. After being privileged to try this dainty a few times, I am inclined to agree with those classical connoisseurs, for a sweeter and tastier shellfish has seldom come from the sea.

Not only can you try this famous viand, but you can gather your own from our southeastern coastal waters. There are three species of Pen Shells that can be found on our shores, and they look enough alike to be described together. All three live in large, fragile shells usually 5 to 11 inches long, although 12-inch specimens are not rare. In outline they are triangular, somewhat narrowly fan-shaped, with approximately equal valves that gape at the wide end. A glance at the illustration will serve to identify any that you find as one of the

Pen Shells, and if you fail to determine its exact species, it doesn't matter, for all three species are equally edible.

The Pen Shell so highly appreciated by the ancients was the *Pinna nobilis*, a Mediterranean species not found on our shores. The only true *Pinna* of the western Atlantic is the Amber Pen Shell, *Pinna carnea*, and it almost missed our continent, being found only in southeastern Florida, although it is plentiful in the Bahamas and West Indies. It is from 5 to 9 inches long, with a central radial ridge in the middle of the valve that is prominent at the narrow or hinge end and plays out before reaching the margin of the shell. Some specimens have 10 radial rows of scale-like spines; on others these are absent. The color varies from a light orange to translucent amber.

The Stiff Pen Shell, or Prickly Pen Shell, *Atrina rigida*, is moderately thick-shelled and the largest of the three local species. It has a dark-brown shell decorated with 15 to 25 rows of triangular scales that sometimes extend into tubular spines. It is found from the Carolinas to Florida, and through the West Indies.

The Saw-Toothed Pen Shell, *Atrina serrata*, closely resembles the above species, except that it is thinner-shelled and lighter in color. The outside of the shell is sculptured with 30 to 50 finely serrated radial ribs. It is found from the Carolinas to Florida, and around the Gulf Coast to Texas.

The Pen Shells seem to avoid both pure sand and pure mud, so they are usually located on bottoms composed of muddy sand or sandy mud. They bury themselves, small end down, with just the lips of the expanded ends showing. The exposed edges of the shell are quite sharp, and have been responsible for many a nasty cut when stepped upon by barefoot bathers. I have located Pen Shells by wading about with sneakers on my feet and feeling for them through the pliable soles. In clear, calm water one can learn to spot the inconspicuous upper edges of the protruding shell. In many places where a wader would come back empty-handed, a sharp-eyed diver equipped with scuba gear or even just a face mask will often be able to bring in a fine catch of Pen Shells. When you locate a Pen Shell, work your hand down around it until you can grasp the narrow part, and it will pull away fairly easily.

Like the scallop, the Pen Shell has a single comparatively huge muscle. It may be that you have already eaten Pen Shells all unknowingly, for these large muscles often appear on Southern markets

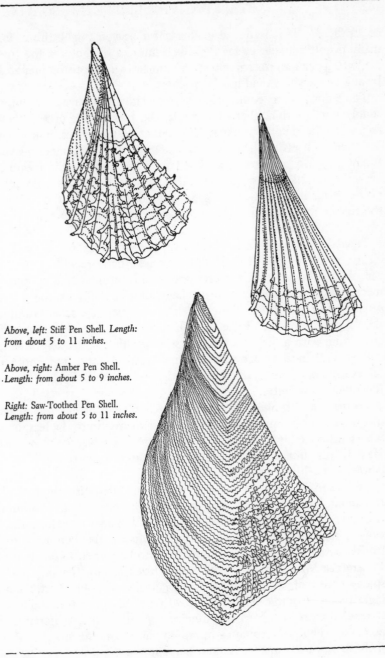

Above, left: Stiff Pen Shell. *Length: from about 5 to 11 inches.*

Above, right: Amber Pen Shell. *.Length: from about 5 to 9 inches.*

Right: Saw-Toothed Pen Shell. *Length: from about 5 to 11 inches.*

as "scallops." This is not as bad a cheat as you might think, for many people actually prefer Pen Shell muscles to those of the true scallops. They are exceedingly sweet and tasty, and can be prepared by any recipe you would use for scallops.

To remove these dainty tidbits from the fresh, living mollusk, simply insert a thin, sharp knife into the gaping lips of the shell and sever the muscle where it joins either of the valves. The shell will then open easily; the muscle can be severed from its moorings on the other valve, and it can be lifted from the body of the creature with the fingers. Although the muscle is the only part ordinarily eaten, this is really a waste of good seafood, for, as with the scallop, the rest of the creature is composed of better meat than some that comes to our table.

To make Scalloped Pen Shells, save the liquor as you open the creatures, remove the "beard" or byssus (the hairy ropelike twist of threads that held the mollusk in place on the bottom), and chop the meat, muscles and all. Scrub as many of the shells as you want servings. In the bottom of each shell put a layer of cracker crumbs, then a generous layer of the Pen Shell meat. Add 2 tablespoons of cream, dot with a little butter, and season with a dash each of salt, pepper, and nutmeg. Cover with fine bread crumbs, add another tablespoon of cream, and dot the top with butter. Bake at 350° for about 25 minutes.

To make a Pen Shell Stew, cook 1 pint of Pen Shell meat in its own liquor for 10 minutes, then add 1 quart rich milk, ¼ teaspoon salt, a little freshly ground black pepper and 4 tablespoons butter. Heat to just under the boiling point, for if it boils the milk will curdle. Serve hot with round pilot biscuits.

The beard or byssus of the Pen Shell is an interesting product of this mollusk, and its use has an interesting history, for it was from this "sea wool" that the ancients spun the thread with which they wove a famous cloth of gold. Like the Mussel, the Pen Shell extrudes these threads to serve as an anchor to keep itself in one place. Where the Mussel spins a very coarse, black byssus, the Pen Shell spins a cluster of golden threads finer than silk. One old report, seemingly trying to impress its readers with the value of this mollusk, states that even though *Pinnas* were so high-priced in Mediterranean ports that poor folk couldn't afford to eat them, the fiber of this shellfish was even more valuable than the meat.

It is said that out of this the fishermen's wives could knit a cloth of gold so fine that a pair of gloves made of it could be folded into a walnut shell, and a large stole or shawl of this material could easily pass through a wedding ring.

Finally, the Pen Shells are closely related to the pearl oysters, and besides yielding sweet meat and golden silk, one rarely—very rarely —finds black pearls of considerable value in their flesh. Surely all this is enough to arouse your interest in this unusual shellfish and send you out looking for this fascinating mollusk.

18. Angel Wings, Fallen Angels, and Rough Piddocks

(*Pholadidae*, or Boring Clam family)

THIS family contains some of the most beautiful and delicious of all clams, and believe me, they have to be good to repay the arduous labor that is expended in collecting them. Unlike other clams that seek out soft sand and mud as a habitat, these clams reside in very hard clay or even soft rock. Other members of this family are able to drill holes in very hard rock or concrete piling, but these ambitious workers will not even be discussed. I love Boring Clams, but I dislike hard work, and not even my enthusiasm for unusual seafood will induce me to chisel it from solid rock.

The Angel Wing, *Cyrtopleura costata*, is better dressed than almost any other bivalve. The beautiful shell is 4 to 8 inches long, and if you look at the two valves spread open, you will never wonder why they are called Angel Wings. Appropriately, they are snow white, sometimes with tinges of "angel pink" inside. The living creature inside is yellowish in color, with reddish-brown dots about the ends of the siphons. This clam is found from Massachusetts to Central America, but is far from being abundant in the northern part of its range. Rachel Carson reports digging Angel Wings from hard mud banks in Buzzard's Bay, and I have taken a few from balls of clay cast ashore during a storm at Ocean City, New Jersey, but it is chiefly a warm-water creature and becomes more common as one travels south.

While doing research on this clam, I stumbled onto a 30-year-old, out-of-print book that told of a fisherman's finding a bed of Angel Wings, many years ago, near a creek mouth in southern New Jersey. From the description of the terrain, I thought I recognized a place where I had often been fishing. The next time I was there I walked along the shore, and sure enough, I began finding broken bits of Angel Wing shell. When the tide was at its lowest, I waded out on a peaty bottom and began digging around the interesting holes I could see through the water. With a pick and a narrow spade I finally discovered six live Angel Wings—found by following a thirty-year-old clue.

The two beautiful winglike valves cannot close completely at the ends, making it very easy to slip a thin knife inside this clam to sever the adductor muscle. The valves are then very easy to pull open. The adductor muscles are quite large, so I separated them out to save for frying, after the manner of scallops. The rest of the meat, which includes the strong foot and the speckled siphon, was finely ground and made into a chowder (page 21). I fried the 6 muscles with a few anise seeds. As one would expect from an Angel Wing, the taste was heavenly. No wonder this is the favorite shellfish around the larger West Indian Islands.

I would not hesitate to substitute Angel Wings in almost any of the clam recipes given in this book. Just imagine the cries of delight that would greet stuffed Angel Wings, served right in those beautiful shells. The adductor muscles can be used in any way you would prepare scallops, and from the one taste I had, I would say that they are at least the equal of that excellent seafood.

The Fallen Angel, *Barnea truncata*, also called the Truncated Borer, has a family resemblance to the Angel Wing, but is considerably shorter, attaining a maximum length of only about 3 inches. The shell is snow white to gray, and gapes widely at both ends; the posterior end is squarish with a cut-off appearance. The radiating ribs that give these shells their winglike appearance are more conspicuous on the anterior end of the Fallen Angel, the posterior half of the shell being comparatively smooth. The shell is thin and somewhat fragile; one seldom finds unbroken shells of the Fallen Angel along the beach. It ranges from Maine to Mexico, and is often abundant in hard clays, soft rocks, or even old timbers that have lodged between the tides.

While visiting on the Elizabeth Islands, which lie between Buzzard's Bay and Vineyard Sound, just off Wood's Hole, Massachusetts, I set out one day determined to find an Angel Wing. There was probably some one-upmanship in my motive, as I wanted to equal Rachel Carson's record of finding one this far north. At the ebb of an exceptionally low tide, I found a stiff clay bottom riddled with holes that fairly shouted, "Angel Wings!" Made eager by anticipated success, I began digging furiously, but alas, after a half-day's laborious digging I had nothing but several dozen Fallen Angels. When I returned to the home where we were staying, my wife remarked that it was somehow typical of me to go out seeking angels' wings and return with fallen angels. However, when I had removed the shells, ground the meat in a food chopper, mounded it into scallop shells picked up on the beach, covered each mound with a slice of bacon, and cooked it all to a fragrant goodness under the broiler, everyone agreed that these angels had not fallen to a very low estate.

The only other Eastern Boring Clam with which I have had experience is the Rough Piddock, *Zirfaea crispata*, which is only about 2 inches long. It bores in very hard clay or soft rock and is fairly common in cold-water areas, although its total range is from Labrador to North Carolina. The queerest thing about this clam is that each valve of its shell is divided into two approximately equal areas by a radial groove that runs directly from the hinge section to the margin of the shell, about midway on the shell. The sculpturing on the two areas is vastly different. The anterior half of the shell is covered with radial ridges or wrinkles, terminating in rasplike teeth at the margin, while the posterior end of each valve is decorated only with concentric growth lines. It looks almost as if two dissimilar clams had been grafted together at the middle. The ends of the shells gape widely, but the margins converge and touch near the middle of the clam.

Rough Piddocks are small and very laborious to dig, but they are almost worth it. As with other Boring Clams, there is a large por-

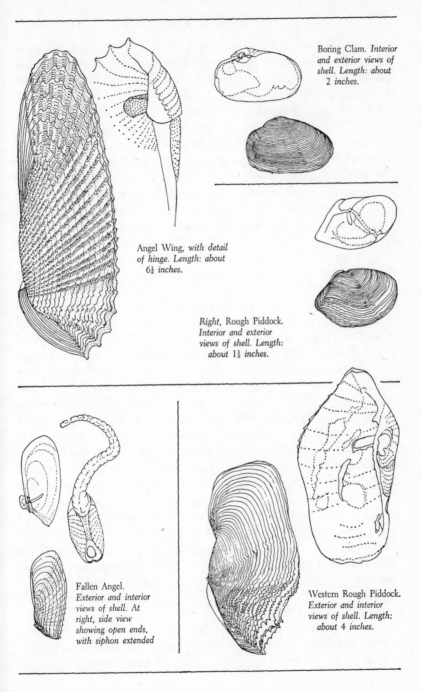

Boring Clam. *Interior and exterior views of shell. Length: about 2 inches.*

Angel Wing, *with detail of hinge. Length: about 6½ inches.*

Right, Rough Piddock. *Interior and exterior views of shell. Length: about 1½ inches.*

Fallen Angel. *Exterior and interior views of shell. At right, side view showing open ends, with siphon extended*

Western Rough Piddock. *Exterior and interior views of shell. Length: about 4 inches.*

tion of meat relative to the length of the shell, since the clam bulges over. Because of the gaping ends they are very easy to clean, and when the meat is run through a food chopper it makes delicious chowders, patties, or scalloped clams.

Westerners will be glad to learn that the Boring Clams of the Pacific Coast are even better than those along the Atlantic. The Western Rough Piddock, *Zirfaea pilsbryi*, is closely related to its Atlantic namesake, but is a giant when compared to its Eastern relative, attaining a shell length of 4½ inches. Merely to state the length of the shell doesn't begin to describe the size of this clam, for the shell is far too small ever to contain the whole animal at one time. The rough, compound siphon is 1 inch or more in diameter, and often a foot long when extended. The Western Rough Piddock bores a permanent burrow in stiff clay where swift tidal currents keep its home swept free of sand and debris. It is found from the Bering Sea to San Diego.

As with other borers, the Rough Piddock's shell gapes at both ends, and the anterior end is roughened with coarse grinding teeth. It was long a matter of controversy among scientists about how the Boring Clams could excavate holes in substances harder than their own shells. Some maintained that the creature must secrete some chemical that softened the rocks, but it seems to me that this theory raises a larger problem than it solves. What kind of chemical would soften the lime-bearing rocks in which these clams dig and not soften the lime-bearing shell of the clam? Finally one scientist, Dr. G. E. MacGinite, settled the question, at least for the species under discussion, by putting one in a glass tube and observing its excavating technique. He found that it protruded its foot and made the end of it into a suction disk that enabled it to obtain a tight hold on the smooth bottom of its burrow. Then it works its shell up and down, while gradually rotating it, making one revolution in about seventy minutes. This gradually wears away minute amounts of the soft rock or rocklike clay, and the bits worn away are expelled through the siphon. Of course this abrasion also wears away the grinding teeth on the clam's shell, but these are constantly being renewed by growth. Throughout its life of seven or eight years, the Piddock keeps boring and enlarging its home to accommodate its constantly increasing size. It seems a poor sort of life, almost as poor as that lived by human beings who slave at detested jobs for years in order to pay off the mortgages on their own burrows, which they call suburban split-level homes.

The Western Piddocks, although their gaping shells keep them off the market, are highly appreciated by local populations in many areas along the Pacific coast. They are dug from hard-clay shores and reefs with picks or heavy bars, and the chopped meat makes wonderful Clam Fritters (page 56). It is also good as Stuffed Clams (page 21) or Clam Chowder made by any of the chowder recipes given in this book.

Another clam that deserves our attention is the *Platyodon cancellatus*, a relative of the Soft Clam, *Mya arenaria*. *Cancellatus* seems to have no common name except "Boring Clam." These clams literally riddle some clay shores on the Pacific, and I once removed twenty-one of them from about a cubic foot of clay bank that I managed to pry off with a bar. Where they are this thick, they are worth going for.

19. Clamming on the West Coast

EXPERIENCE is the best teacher, and most of the lore and nearly all of the recipes found in this book were gleaned from local informants. This unwritten folklore or "knowledge never learned at school" is invaluable, and contains much information that cannot be found in books, but one must learn to use it wisely, for it is often a strange compound of wisdom, fact, superstition, prejudice, and just plain rot. This is easily seen when one compares the food lore of one locality with that of another. The Blue Mussel is eagerly sought and even carefully cultivated in France, where it competes with the oyster as a delicacy, but in New England, acres of mussels of this same species go begging. The Periwinkle commands a good price in England, yet on our shores it is almost totally ignored. On the other hand, the Venus Clams (our Quahogs or Littlenecks) are considered one of the finest seafoods on our coast, but in France and England they are used only as fish bait and are not considered edible. The octopus and squid are luxury foods around the Mediterranean, in Japan, and in the Hawaiian Islands, but along our coasts they are only eaten by expatriates from those countries, and to most people they are objects of disgust. Examples of this kind could be given by hundreds. Whether or not a food is eaten in a certain locality has little to do

with its intrinsic worth. In our food preferences, even the most civilized and reasonable of us are still largely ruled by habit, tradition,
superstition, and prejudice.

When I give instructions for the use of some perfectly edible but
uneaten life-form along our East Coast, I suppose I am writing mainly
for inland people and Westerners who visit the Atlantic shores, and
when I write of West Coast seafoods I'm sure my instructions will
find more use among Easterners who vacation along the Pacific.
Those who have lived their whole lives near one section of seacoast
have long since allowed habit and custom to dictate which of the
local seafoods they will eat and which ones they will not, and most
of them are not about to change or try anything new. On the other
hand, one who visits a new place with new eating customs is forced
to change his habits somewhat, and since all the foods are new, it
makes little difference to him whether the new food is in line with
local food prejudices or not.

It would seem logical to assume that the best way to learn about
forageable seafoods in a new area would be to have some experienced
local person point out the various life-forms and explain their uses.
It is logical, but in practice it often works very poorly. Local information should be taken with a very large grain of wholesome, mineral-
filled sea-salt, and at all times should be supplemented and irradiated
by knowledge and experience.

The first time I explored the beaches of Puget Sound, many years
ago, I took along a young fellow who had been recommended to me
as an expert forager and seafood enthusiast who knew all about the
life on local beaches. I soon discovered that he really had a tremendous store of misinformation and a very broad firsthand unfamiliarity
with the facts. However, he was a pleasant companion, and he did
know the local shore and the tides and where the different creatures
could be found, although he was a bit hazy about their identification
and uses.

It was an ideal day for studying and collecting littoral life for there
was an exceptionally low minus tide, baring areas of mud flat and
beach that were seldom seen. We made our way around the base of
a cliff, walking on land that was above water only a few times per
year, and consequently unspoiled, although it was not far from a large
population center. Huge rocks that had fallen from the cliff in ages
past stood like scattered houses on the flat, each covered with a lux-

uriant growth of plant and animal life. There were thick clusters of Blue Mussels, my old friend *Mytilis edulis*, but my guide told me that all mussels were poisonous. While it is true that mussels found on the outer West Coast are likely to contain poison in summer and early autumn, this was March and these mussels were in protected, inland tidewater. They were really wholesome and perfectly delicious, as I subsequently proved to my own satisfaction by gathering and eating large quantities of them.

The mussels on these rocks had so intermixed their byssal threads that one could pull off whole clusters of them at one time. When I pulled off such a cluster, under it there were half a dozen little Blennies, about 6 inches long, that flapped excitedly for cover, but I captured them. My companion said these were "eels," and that they made good fish bait but weren't good to eat. On questioning, I found that this bit of information had come from that mysterious "they" who are the source of so much of this world's "knowledge." "They" had said these were eels and that eels were inedible, and it had never occurred to him to test these assertions by further research or experiment. Actually Blennies are fine fare, and I was overjoyed to discover such an abundant and easily-acquired supply of them.

As we crossed a flat of gooey mud, I shoved in my spade and turned up some Bent-Nose Clams, *Macoma nasuta*, but my mentor told me these also were inedible "because they are full of mud." This clam does need a bit of processing to rid it of mud, but when this is done it has very delicious flesh. At one time this was a popular clam in the San Francisco markets, and it must have been a favorite of the ancient Indians along this coast, to judge from the number of its shells that are found in the mighty shell mounds or kitchen middens this seafood-loving people left behind them.

Where there were mixed sand and cobbles, many specimens of the Basket Cockle, *Clinocardium nuttallii*, lay half uncovered. One could have gathered a bushel of them without doing any digging, but my companion was uninterested; he said these were not good to eat, because they were cockles and not clams. I had seen this same species of cockle bringing high prices on the San Francisco market, for they are scarce in California, but here truckloads of them lay unused. This cockle is a bit tough, but no more so than the highly prized Eastern Quahog, and like the Quahog it makes superb chowder and delicious Stuffed Clams.

On a rock, still being lapped by the tide, was a great cluster of small Native Oysters, *Ostrea lurida*, but my guide also rejected these because they were "Indian Oysters and not real oysters." Not many miles from where we stood, there were huge oyster farms where this same species was raised commercially in cultivated beds and the shucked product sold at premium prices on local markets. On inquiry I found that my guide was very fond of these little oysters when he paid outrageous prices for them as "Olympia Oysters," not realizing that these were the same plentiful species that he spurned on the beach as "Indian Oysters."

At the base of another wave-washed rock, I saw a 4-inch shell just under the receding water with its values half open. As I bent to examine it more closely, the valves suddenly snapped together, shooting a jet of water in my face. In revenge I turned the rock over and saw several more similar shells attached to its bottom, their shape conforming to that of the rock to which they were firmly attached. My local informant told me these were "Rock Oysters," and said he had never heard of anyone eating them. This "Rock Oyster" is not an oyster at all; it is a scallop, *Hinnites giganteus*, and the young of this species are free-swimming and look much like other scallops, but they soon settle on some rock and then grow to fit the foundation. As in other scallops, the adductor muscle of this sedentary species is a toothsome morsel.

On the shores of a swiftly running tidal channel a little farther on was a bed of almost rock-hard clay pitted with holes about 1½ inches across. As we stepped near one, a little squirt would signal that the creature below was pulling in its large siphon. Here I received a lecture in "unnatural history." These receding siphons, my guide assured me, belonged to Geoducks, but, he made sure to inform me, Geoducks were only edible when taken from sandy shores. When they settled in clay, he explained, only their "necks" grew to full size, while their bodies remained small, misshapen, and inedible. His "Geoducks" were really Rough Piddocks or Boring Clams, *Zirfaea pilsbryi*, and they are perfectly edible, even delicious, although one who digs enough of them for a meal from the hard clay in which they live has usually earned his dinner.

On a flat of sandy mud there were tremendous squirts of water as we passed, shooting three feet or more into the air, making the little squirts of the Boring Clams look feeble by comparison. I insisted on digging out some of these creatures, and after sinking a hole more than two feet deep, I turned out a huge clam with a shell about 7 inches long that still wasn't large enough to fit its overgrown body, and a great brown, rough-looking siphon hanging from one end. The whole clam must have weighed more than 3 pounds. My guide tolerantly informed me that I had wasted my time digging out this useless creature, for, he said, this was a Horse Clam and was not edible, because it lived on beach crabs and always had some half-digested crabs still alive in its stomach. I was especially amused at the reasons this lad gave for not eating the edible shellfish we were finding in such abundance; I'm afraid most of us have perfectly logical and convincing reasons for our illogical and unreasonable prejudices. The huge clam I was holding had been correctly identified, for it is known variously as the Horse Clam, Gaper, Big-Neck, Rubberneck, Otter Shell, Summer Clam, and Great Washington Clam. After wading through such an appalling list of common names, one is almost ready to learn that its scientific appellation is *Schizothaerus nuttallii*. Contrary to the local information I was receiving, it is edible right down to the end of its big, ugly siphon, when properly prepared. Like other clams, it lives on microscopic food particles taken from the water pumped in by its siphon, and the little "half-digested" crabs that bothered my companion are the commensal Pea Crabs of several species that pass their whole lives in the mantle cavity of this clam, snitching food from the water the clam pumps past them while escaping the hazards of existence that characterize life in the open sea. I decided not to outrage my informant by telling him that I intended to eat not only these "inedible clams," but also the very crabs which he insisted made the whole clam inedible.

I have been unfair to this boy, for unwittingly he did give me some solid information, and he finally led me to as fine a bed of Washington Butter Clams as I have ever seen and gave me many useful hints on how to prepare them. However, such an experiment in using local information often leaves me with a sense of frustration. One cannot teach people a subject about which they think they already know all that is to be known. The presentation of evidence means nothing, for many of them will still trust their own folklore in the face of all evidence to the contrary. The Horse Clam that is disdained about Puget Sound is so eagerly sought in California that game laws had to be passed to protect it. I have come to the conclusion that the old cracker-barrel philosopher of my boyhood, Josh Billings, was right when he said, "The worst kind of ignorance ain't so much not knowin' things as it is knowin' so many things that ain't so."

THE WASHINGTON CLAM, BUTTER CLAM OR MONEY CLAM
Saxidomus giganteus and *S. nuttallii*

The local guide execrated above deserves credit for finally leading me to a sandspit that only came above water during the lowest minus tides, where the Butter Clams lay crowded together a few inches under the surface. Perhaps it was the availability of an abundant supply of this outstandingly good species that really kept him from sampling the other kinds of seafood all about him, for this little clam is good enough to make many other kinds seem inedible by comparison.

Other clams may be as good as or better than the Butter Clam, but none is so versatile. Cherrystones are better raw, Razors better fried, Coquinas and Bean Clams make better nectar, Quahogs are better stuffed, and Surf Clam adductor muscles make better chowder, but the Butter Clam can be prepared in any of these ways, and it will rival the best. You can use Butter Clams in any clam recipe you can find, and it will still be delicious. To paraphrase what was once said about the wild strawberry, "Undoubtedly God could have made a better clam, but undoubtedly God never did."

The species we were gathering was *Saxidomus giganteus*, a comparatively small clam, almost circular in outline and about 3 inches in diameter. The plump little shell is marked with no radial lines, but shows pronounced concentric lines of growth. It is found from as far up in Alaska as you are likely to go, to Monterey, California, and is the commonest clam eaten in the Puget Sound area. The very similar, but slightly larger, *Saxidomus nuttallii* is found from Humboldt Bay to Lower California. This species sometimes achieves a diameter of 6 inches, and 5-inch specimens are not uncommon. Inside the shell, it has purplish markings near the siphon end that are missing in *giganteus*. Both kinds are called Butter Clams, or Money Clams, the latter name arising from a tradition that the California Indians used the round shells as a medium of exchange. The two species are so much alike in flavor and uses that where their ranges overlap in northern California, clam diggers do not distinguish between them, and they are all sold together on the market. You will note that it is the smaller of these two species that is called *giganteus,* thus proving that even scientific names can be misleading.

THE BENT-NOSE CLAM
Macoma nasuta

The Bent-Nose Clam that had been dismissed as being full of mud could hardly be otherwise in the environment in which it lives. It is found in mud flats, and the softer the mud, the better. It can live in softer mud than any other clam; one can often just reach down into the ooze and pick them up. It can also tolerate water so stale that other species die out, so this is often the only clam found along sloughs and lagoons that are only connected with the ocean at highest tides. This clam can probably be found on every suitable mud flat from Alaska to the Gulf of California. It is not a large clam, attaining a maximum length of only about 3 inches, and it is rather thin. The shell is white, but in life it is usually covered with a gray periostracum. The siphon end of the shell is elongated and bent to the right; since the clam lies on its left side, its bent nose probably helps to point its siphon toward the surface.

Despite its small size and muddy habitat, this can be quite a delicious clam when properly processed. It was once a favorite on the San Francisco markets, being gathered and freed of mud by Chinese clam diggers. In the old days, these Sons of the Celestial Empire often suffered from job discrimination, so many of them became clam diggers. They built shacks along the tidal flats of San Francisco Bay and went into business for themselves. Such a frugal people would not waste even a muddy little clam with a bent nose, so they developed a method of freeing this clam's viscera of mud, thus making them acceptable on the local market.

These Chinese built wooden vats or tanks with a false bottom of slats raised above the real bottom. The Bent-Nose Clams were piled on the slats, and the vats were filled with clean seawater. At each high tide the stale water was flushed from the vats and they were refilled with clean seawater. After three days, the clams had evacuated all their contained mud and were completely free of any muddy flavor.

A seaside vacationer cannot be expected to construct and tend such a complicated apparatus merely to enjoy a few meals of Bent-Nose

Clams, but the significant fact here is that when these clams are suspended in clean seawater for three days they become clean and sweet. My first attempt to take advantage of this fact ended disastrously. I tied about 2 gallons of Bent-Noses in an old piece of fishnet and hung it in the water from the end of a pier. Three days later, when I was anticipating a meal of super-cleaned clams, I pulled up the rope only to find nothing but a few frayed pieces of netting on the end of it. A shark or ray had happened along, and finding this nice supply of its favorite food, had made short work of my fragile net bag.

To outwit these predators, I found a metal milk-bottle crate and covered it with heavy, galvanized, ½-inch-mesh hardware cloth. This worked to perfection, and Bent-Noses kept in this underwater cage for three days were as sweet and delicious as Butter Clams. In fact, I later learned that Butter Clams and many other species could be improved and entirely freed of sand, grit, and mud by this same method.

THE WHITE SAND CLAM
Macoma secta

Also known as the Giant Macoma, this close relative of the Bent-Nose is found in sandy flats from British Columbia to the Gulf of California. It looks like the Bent-Nose but is somewhat larger, attaining a length of 4 inches, and it doesn't have a bent nose. Although occurring over much the same range, because of their different habitats these two similar clams seldom meet, except on my dining table.

The White Sand Clam lives in pure-sand flats about 1 to 1½ feet below the surface. This sounds deep enough to mean hard work, but they are often so thick that one can dig a hole about a yard square and gather all the clams one could want from around its caving banks. This clam is delicious in flavor, but is seldom used for food because when first collected, it, too, is just too gritty to eat. After being suspended as the Bent-Noses were for three days in clean seawater in the wire cage, they seemed to me to equal almost any clam.

One would think that these two Macomas, one adapted to sand and another to mud, would monopolize every tidal flat within their range, but there is a broad spectrum of sandy mud and muddy sand where neither is found. These in-between spots are, however, often loaded with Butter Clams, Horse Clams, Jackknives, Cockles, and other edible species. I can't imagine a clever forager having to travel far on the West Coast to find edible clams of one species or another.

When I say that these two *Macomas* are as good as other species, I don't mean to imply that they taste exactly like other clams, for they have a very good but distinct flavor, not only from other clams, but even one from the other. However, this flavor is definitely within the clam class, and you can prepare either or both *Macomas* according to almost any clam recipe you can find. They are excellent just steamed and eaten hot with Melted-Butter Sauce (page 21); they can be opened raw, dipped in batter, and fried according to the directions given for small Razor Clams, or they can be made into a high-class chowder. The following two recipes are given, not because they could not be used with other species, but because I first tried them with *Macomas*.

To make Clam Spread Canapés, use only clams that have been suspended in seawater until they are thoroughly cleaned of mud and sand. Put 2 quarts of clams and ½ cup of any good, dry, white wine in a kettle with a tight-fitting cover, place over high heat, and steam for 10 minutes. Remove the clam meat and chop fine. Mix chopped clams with 3 ounces cream cheese, adding enough of the winy broth from the steaming-kettle to make it soft and spreadable. Add 1 tablespoon finely minced onion and 1 tablespoon mayonnaise. Season with salt and black pepper to taste, mix well once again and spread on toast rounds, decorating each canapé with a sprinkle of paprika and a little sprig of emerald-green samphire (page 258). These are excellent as hors d'oeuvres, or go very well with a buffet luncheon or smorgasbord. Of course these *could* be made of canned clams and garnished with ordinary parsley, and they would be good, but then you wouldn't have that wonderful and altogether different feeling toward the food, which comes only when you have garnered the main ingredients with your own two hands.

French Fried Clam Bites could be made of any species of edible clam, but I can testify that they are delicious when made of either of the *Macomas*. Steam 2 quarts of internally clean clams for 10 minutes, this time using ½ cup of boiling water instead of wine to start the steaming. Remove clams from shells and mince fine. Separate an egg and beat the yolk until it is fluffy and lemon-colored, add 1 tablespoon melted butter, ½ cup flour, and ¼ cup clam broth, then beat

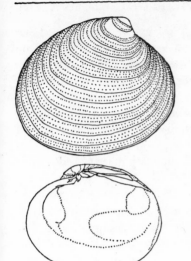

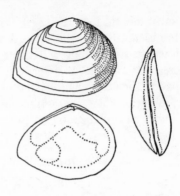

Left, Butter Clam. Exterior and interior views of shell. Length: about 3 to 5 inches. Above, Bent-Nose Clam. Maximum length: 3 inches.

White Sand Clam. Exterior and interior views of shell. Length: about 4 inches.

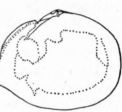

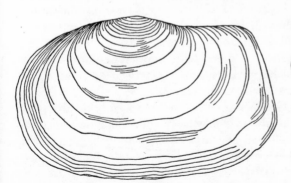

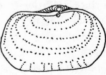

Geoduck. Left, exterior view. Right, interior of shell.

until smooth. Beat the egg white until it is very stiff, then fold it into
the batter. Stir in the minced clams and let everything stand for at
least an hour. This will cause the batter to thicken and become
creamily smooth. Stir again to mix the clams evenly through the
batter, then drop by teaspoonfuls into hot fat, 375°, and fry for about
5 minutes, when they will be golden brown. Serve as part of a seafood
dinner, at buffet luncheons, or on toothpicks as hors d'oeuvres.

HUNTING THE WILD GEODUCK
Panope generosa

The Geoduck is not a duck; it is not even a waterfowl, nor is it a
bird of any kind. It is an outsize clam, probably the largest intertidal
bivalve in the world. From southern Alaska to northern California,
the Geoduck lies deeply buried beneath that narrow strip of shore
that only comes above water during the very lowest tides. The beachcomber who is fortunate enough to bag one of these huge clams will
find himself in possession of a prize piece of meat, large enough to
cut good-sized steaks from it and have enough left over for a first-class
chowder. The shell may be 7 inches long, but that doesn't begin to
describe the size of a Geoduck, for this poorly planned creature has
a shell much too small to fit it and bulges over in all directions.

Some local clam diggers will tell you that the Geoduck can burrow
swiftly through the sand, but this is not true. It lies snug and secure
about three feet beneath the surface and bets that you will give up
trying to dig through the caving, flowing, wet sand before you reach
it. To survive at this depth, the Geoduck must have a pipeline to the
surface in order to draw in food and expel wastes, and a pipeline is
just what it has, a huge compound siphon, or two-channeled tube,
that will extend three feet or more. Laymen often call this siphon a
"neck," but it would be more accurately described as a tail for it
originates at the posterior end of the clam.

Hunting the wild Geoduck is eerie fun and so fantastic an operation that the neophyte always thinks he is being kidded when someone explains how it is done. If you are extremely lucky, you might
find a Geoduck at the ebb of an ordinary spring tide, but to be sure
of bringing home your quarry, you must seek it during those extremely
low tides that occur only in the spring and fall. There is something
weird about hunting a creature that can successfully be found on only

two days of the year, and it adds to the magical quality that those two days must be computed from a lunar rather than from an ordinary solar calendar. Along in late October or early November, there comes the full of what some tribes of Indians call the Mad Moon. As this moon waxes and grows round, the tides leap higher on the beach and ebb farther out than they will for another half-year. The very lowest of the autumn tides will occur approximately two days after the full of the Mad Moon, and in the section where Geoducks will be found, the lowest of the diurnal tides will come in the middle of the night. I know it sounds like a psychotic sorcerer's formula to say the Geoduck must be sought at midnight, just two days after the full of the Mad Moon, but it happens to be a sober fact.

It takes two hunters working together to bag a Geoduck, one a neck man and the other a shovel man. Walk along that mysterious strip of moonlit beach that hasn't been above water since last April, and sooner or later you will see a squirt like a little fountain. This is the Geoduck expelling the water from his firehose of a siphon, preparatory to retracting it, for even the Geoduck, with only a very diffuse nervous system and nothing that could be called a brain, knows enough not to stick out its neck when a clam digger is around. The split second such a squirt is detected, the neck man fairly dives onto it and gets a firm grip around the retreating siphon with his hand. But don't try to keep that siphon from retracting, for the Geoduck has powerful muscles for pulling his neck in, and if you insist on holding the end of that siphon at the surface, he will just keep tugging until suddenly the neck will snap in two. The thing to do is to keep a firm grip on the neck but allow your hand to follow it down through the sand. You will probably end up lying flat on the wet beach with your arm in up to the shoulder, but finally you will feel your whole hand sink right into the soft flesh of the Geoduck. After that, your job is just to lie there and hold on while your partner excavates a hole about your arm so the creature can be lifted out.

I once introduced a friend to this sport, and at first I was always the neck man. After a bit my partner began thinking I had too easy a thing of it, just lying down on the job and letting him do all the hard digging. Finally he insisted that I take the shovel while he tried his hand at necking. He managed to get a firm hold on the next Geoduck that squirted, and followed him down successfully, but when his hand sank into that soft, gooey flesh he panicked and jerked his hand away. When I asked him why, he said, "I was afraid he would bite me."

"Now look," I sagely said, "you have seen Geoducks, cleaned them, and eaten them. Did you ever see one with teeth?"

"No," he admitted, "but some day there will be a first one, and it will be just my luck to grab him."

I can almost hear my readers exclaiming, "Surely there must be an easier way of digging these clams!" Yes, I know at least two easier ways, but they are both illegal, and I refuse to be an accessory before the fact of your crimes. The daily bag limit in the State of Washington is two Geoducks per hunter, or four for a team of two, and that is plenty. One of these huge clams will furnish cutlets enough for a meal for a large family, and a single one of those exaggerated siphons can be ground up to make a large kettle of chowder. You won't need many recipes for preparing Geoducks for you will meet this delicacy on your table only about twice a year if you keep within the game laws.

To make Geoduck Cutlets, first sever the adductor muscles and remove the undersized shell. Then grasp the huge siphon in one hand and, while keeping tension on it, cut around the viscera with a sharp knife until the tube comes away in your hand. Cut the sandy viscera away and discard it. Wash the body under running water to remove the clinging sand, then slice it into cutlets of a convenient size. Beat 1 egg with a tablespoon of water, dip the cutlets into it, then dredge with fine cracker crumbs and fry to a golden brown on both sides. They are excellent.

The siphon is too tough for frying, but has a fine flavor, so use it to make Geoduck Chowder. Scald it in boiling water, then peel away the brown, outside skin. Grind it in a food chopper or sausage mill, adding half a dozen slices of raw bacon as you grind. Put the ground clam-bacon mixture into a frying pan with one fine-chopped onion, and fry until all is lightly browned. Transfer this to a deep stewpan, and add 2 cups of diced potatoes and 1 quart of water or clam juice. Boil until potatoes are done, then pour in 1 quart of milk, reserving ½ cup to blend with the thickening. The thickening is made by stirring into the ½ cup of milk 1 tablespoon flour, 1 teaspoon monosodium glutamate, and ¼ teaspoon freshly ground black pepper. Stir this into the chowder a little at a time, and bring the soup just to a simmer. Never boil Geoduck Chowder after the milk has been added.

Serve very hot with saltines. The Geoduck is delicious to eat and grand fun to catch, so when you are in the Pacific Northwest, make his acquaintance.

THE HORSE CLAM, GAPER, OR OTTER SHELL
Schizothaerus nuttallii

This is the huge clam my local informant rejected "because it lives on beach crabs," which it certainly does not. It sieves tiny particles of food from the water it pumps in and out of its mantle with its siphons, just as other clams do. And the white commensal crabs almost always found in this clam's mantle cavity are not half-digested —their easy, secure life has caused them to degenerate, and their shells never harden, their eyes are useless, and their legs too weak to make any but feeble movements. These freeloaders are of three species. The larger ones are usually *Pinnaxa faba*; there will probably be several *Opisthopus transversus*, and occasionally the much smaller *Pinnaxa littoralis* will be found. Collectively, these commensal crabs are known as Pea Crabs, and each of the three species may be represented by several individuals in one Horse Clam, so a large Horse Clam sometimes supports as many as a dozen of these nonpaying boarders. They apparently do the clam no harm beyond stealing the choicest tidbits from the ample food which he constantly pumps in.

To the gourmet who isn't squeamish about new foods, these little crabs come as a bonus. They make an epicurean dish when sautéed in butter or battered and fried, being the equal of the finest Eastern Soft-Shell Crab. For descriptions of these crabs and directions for preparing them, see page 40.

The Horse Clam is the second largest bivalve of the Pacific intertidal zone, rivaling the Geoduck in size, for it attains a length of 8 inches and a weight of 4 pounds. It is often mistaken for the Geoduck, and whenever you hear someone telling of finding great numbers of Geoducks on an ordinary low tide, rest assured that their Geoducks were Horse Clams. If calling a clam by another name makes the chowder taste better, I would not disillusion them, for the Horse Clam is eminently edible, with large quantities of sweet, tasty flesh.

Since this large clam is not likely to be mistaken for anything except a Geoduck, it might be well to compare the two. Both are of large size, lie deep beneath the surface, and have unbelievably long, extensible siphons, usually called "necks," but here the resemblance ends. The Geoduck is far too large for its shell and bulges over all around, being unable to close its valves at any point. The Horse Clam is also a bit large for its shell, but bulges over only at the siphonal end, where its shell gapes widely, and it is able to close its valves around the rest of its body. Of course it can never retract its huge siphon completely within its shell, so that must shift for itself. The siphon is protected by a tough, brown skin and has at its tip two horny valves that can be used to close the openings. This is an excellent recognition feature possessed by no other common clam. When seen on the tide flats, these horny lids are usually covered with a gob of mud; whether it is there to serve as camouflage, or because this clam is just plain untidy, I do not know, but I suspect the latter.

Throughout its range from Alaska to San Diego, the Horse Clam is unevenly popular, being prized in some areas and rejected in others. Because of its ill-fitting, gaping shell and protruding siphon, it cannot retain moisture like some other species, and this makes it a poor market clam. Some people, like my local informant, are repelled by the presence of the little white commensal crabs in its mantle cavity, but these really do not harm the meat in any way, and besides, these crabs are good to eat, too. I suspect it is mainly this clam's appearance that militates against its popularity, for it is certainly no beauty. The rough, gray shell is covered in life by a brownish periostracum that is often rubbed off in spots, giving the creature a moth-eaten appearance. The bulging body, usually muddy, protrudes from the gaping siphonal end of the ugly shell, and from this part of the body comes the overgrown compound siphon, rough and brown, usually flecked with mud, and with a horny tip. All these features combine to make the Horse Clam look anything but appetizing. If we had never eaten seafood of any kind, how many of our common shellfish, either crustacean or molluscan, would look edible on first acquaintance? However, those who can overcome their repugnance at this clam's appearance when it is first lifted from its muddy bed are in for a treat. When it comes to the table as hot clam patties, fried to a golden brown and decorated with appropriate garnishes, you will find its appearance vastly improved.

Unlike the Geoduck, which is only found at the very lowest minus tides, the Horse Clam is found on mud-and-sand flats that are uncovered by anything lower than mid-tide. It is the only clam I know that can be located from a distance. Whenever it retracts its huge siphon this clam shoots a regular fountain of water into the air. Other clams do this also, but no other besides the Geoduck has such a volume of water or shoots it to such heights. The Horse Clam turns on this fountain when disturbed, and it happens far too often just as I am stepping over the opening for me to be fully convinced that its aim is accidental. I have been wet to the waist by a single squirt. This clam also apparently pulls in that long siphon frequently, even when not disturbed, for I have stood at one edge of a mud flat and located the richest beds of the largest clams by observing the size and number of the jets seen in the distance.

It is a job to dig them out for they are often 2 to 3 feet below the surface. Locate a large "squirt-hole" (as it is called among clam diggers), and dig down to water-bearing sand. Work your hand down through the yielding medium until you locate the clam, then slip your hand under its shell and work it to the surface. California law allows you to take no more than ten at any one tide, but only a glutton or a vandal would want more than ten of these huge clams at one time. If you must transport the clams far, or intend to keep them for several hours before cleaning them, pack your catch with wet Rockweed to keep the clams from drying out.

I have never tried it, but I feel sure that the Horse Clam could be improved by suspending it in clean seawater, as directed for the Macoma species (page 106), but this is not absolutely necessary. Wash as much sand as possible from the outside of the clam before opening it. These big clams are unsuitable for steaming, and should always be opened raw. Insert a long, thin knife into the open end of the shell and sever the adductor muscles, then pull the clam open. Do this over a container so you can catch the copious juice that flows out.

Remove, but do not discard, the big ugly siphon, for many consider this the finest meat in the Horse Clam. Some people prize these clams only for their siphons and throw the rest of the clam away, but I consider this a criminal waste of some very good food. The siphons should be dipped in boiling water for 10 seconds, then skinned, removing all the brown epidermis and leaving a double tube of white flesh. This should be split open lengthwise, then cut cross-wise, making 4 pieces. Pound each piece with a rolling pin or a mallet, as described for Abalone. When beaten into submission, it can be dipped in beaten egg, dredged in seasoned flour or crumbs, and fried exactly like Abalone. It won't taste like Abalone, and it doesn't need to. It has its own flavor and texture, and these are good enough so that this food doesn't have to imitate anything else.

Decant the clam juice into another container to eliminate the settled sand. Remove all the sandy viscera from the clam meat, then wash the pieces in the clam juice, adding a little water if necessary. Washing the meat in clam juice prevents the really good flavor of the Horse Clam from being diluted. When you are sure the clam meat is free of sand and mud, put it through a food chopper.

To make Clamburgers, mix 2 cups of ground Horse Clam meat with 1 beaten egg and just enough dry bread or cracker crumbs to allow you to shape it into patties with dampened hands. It really needs no seasoning, being naturally salted by the sea. Fry just as you would hamburgers and serve on buttered, toasted rolls.

This ground clam meat also makes excellent Clam Fritters (page 56), Chowder (page 21), or Stuffed Clams (page 21-22). When you are anywhere on the West Coast, find a mud flat and gather yourself some Horse Clams. If a local resident happens along and tells you that Horse Clams are not edible, well—just invite him to share your dinner.

THE PRIZED PISMO CLAM
Tivela stultorum

Had a Californian written this book, there is no doubt that the Pismo Clam would have occupied a very prominent position. When I was in Pismo Beach some years ago, I saw native sons seeking this clam with an avidity and earnestness that gave the operation some of the characteristics of a religious observance. Several small towns, Morro Bay, Pismo Beach, and Oceano, seemed to exist primarily to cater to the communicants of this cult. Tourist facilities and camp grounds were filled with people who had come to spend their vacations in pursuit of the Pismo. Orthodox equipment was offered for sale or rent, for a respectable clammer doesn't disturb the Pismo with just any old makeshift tool. Groups gathered on the beach and street corners to discuss clams and clamming.

The teeming life at the edge of the sea had always fascinated me, and clam digging and allied activities were even then one of my chief recreations, but I had already discovered that my hobby bored most people, as they considered it too tame to be an interesting sport. Imagine my delight at discovering a town full of clamming buffs. Naturally I was tempted to alter other plans and join them, and of course I succumbed. A local shop outfitted me with a rake and drag. The latter is a netted bag about 4 feet long and some 8 inches in diameter, held open at the mouth by being netted around a wooden hoop. It comes complete with a web belt to which the drag fastens with a bridle and a large snap fastener. The snap fastener is a safety feature enabling the clammer to cast off the bag easily if he gets into trouble in the surf. The net bag is long enough so that clams dropped through the wooden hoop (the mouth of the bag), rest on the ocean bottom and are not a dead weight hanging on the clam digger. When the digger moves about, the load of clams is dragged behind, hence the term "drag" that is applied to this apparatus; it serves as a sort of combination anchor and game bag. The rake used on this coast is long-tined and long-handled. A piece of rope the length of the rake handle is tied to the rake head. In use, the tines are pulled through the sand by the rope while the rake is guided by the handle.

I strolled along the beach to see if I could pick up any pointers from returning clammers who were being driven ashore by the current tide. Although all of them were wet from head to toe, they wore old clothes rather than bathing trunks. I was sure that I would rather be in that surf in bathing trunks than in sodden, heavy clothes, but such is the tyranny of style that I decided to dress in that fashion so others would not know that I was a neophyte at this game.

When I arrived on the beach the next morning, a cold fog was blowing in from the sea and the boiling surf looked anything but inviting. As I hesitated, other clammers passed me and plunged boldly in, so I followed. The tide was very low and still going out, but there were deep channels between the shoals on the bottom. The experienced hands seemed to know a path that led through the shallows, but I was several times immersed to my shoulders, and once I had to swim for it. Finally, far out, I waded onto a shoal that was occupied by only one other clam digger. By this time the gray, foggy dawn was breaking.

Although the water was probably less than knee-deep on the average, the waves, at this stage of the tide on the shoal, came in one after another, wetting us to the waist, to the shoulders, or over our heads. I was knocked down by several waves that broke near me. Between waves I was trying to master this new skill. I soon acquired the technique of pulling the rake by the rope with one hand while I guided it and adjusted the angle of the tines with the other hand on the rake handle. I also soon discovered that the Pismo always lies with its hinge toward the sea, presenting its narrowest section to the force of the waves. To avoid straddling them with the tines it is better to rake across the beach parallel with the shore and the waves.

I was amazed to discover that my heavy clothes, even when wet, furnished some warmth and protection from the cool breeze. I began to understand the attraction of this activity. Ordinary clam digging might be poor sport, but this seeking the Pismo in the surf was wildly exhilarating. Not that the clams fight back. They don't need to. The clams merely lie quietly about their own length beneath the sand of the bottom and allow the wild waves above to do their fighting. They were scattered enough to make the finding of each legal-sized clam a minor thrill. At that time the game laws allowed each digger to collect fifteen Pismos of a minimum length of 5 inches at each tide. I had two notches cut on my rake handle, just 5 inches apart, to determine the legality of the clams I discovered.

The fog lifted and the sun came out, revealing other shoals, each with its quota of clam diggers, all about us. The battle with the waves intensified as the tide started back in. My companion, having acquired the bag limit of extra-large clams, offered to guide me through the shallows back to shore. I had only half a dozen clams in my bag, but these averaged nearly two pounds apiece, so I felt that I had all the clams I needed and accepted his offer.

I was staying with some relatives, an old couple who had retired to California, so I offered to cook my clams for a midday dinner. At first those heavy shells, so tightly closed, looked unassailable, but I found that if one gave the clam no prior warning a thin-bladed knife could be deftly slipped between the valves. When the adductor muscles were severed, the clam opened of its own accord. Those heavy, large shells seemed to be made for use as cooking containers, and these suggested Stuffed Clams.

I washed the clams in their own juice to remove any sand, of which there was surprisingly little. Then I put the clam meat through a food chopper, adding a quartered onion, half a green pepper, and a sprig

Above, Acmaea persona. Left,
upper side; center, under side;
right, side view.

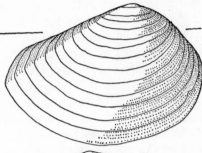

Horse Clam. Exterior and interior
views of shell. Maximum length:
about 8 inches.

Pismo Clam. Exterior and interior
views of shell.

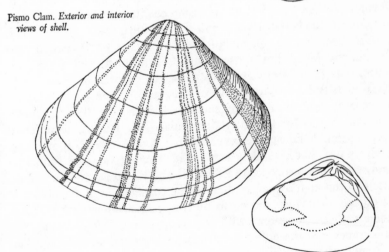

of celery to the grind. I mixed in some bread crumbs, but my stuffing was mostly clam. I mounded this stuffing on 3 large half shells and covered each mound with a slice of bacon. They were cooked in a medium oven until the bacon was crisply done, and then served while it was still sizzling. At the first bite I became a full-fledged convert to the Cult of Pismophagists, and I knew why Californians come year after year to do battle with the waves in order to get a bag of Pismos. It is just a clam, but a clam refined to the absolute ultimate.

The Pismo Clam is so good, and was once so plentiful, that it largely crowded other clams off the California markets, hundreds of thousands of pounds being sold each year. To hear of the way this vast wealth of clams was squandered is enough to make one weep. There are stories of local farmers gathering these clams, by the wagon-load, by opening furrows across the beach at ebb tide with horse-drawn turning plows. Some of the clams gathered in this way were sold on the market, but many a wagonload went back to the farms to fatten the pigs and chickens. The Pismo Clam grows too slowly to endure such intensive exploitation for very long. It takes each clam from four to seven years to reach the present legal minimum of 5 inches in length. A clam 7 inches across, the long way, may be as much as fifteen years old. Despite strict enforcement of drastic con-servation measures, the supply of these excellent clams continues to dwindle. When I was last in the Pismo country, it was almost impos-sible to find a clam of legal size in the intertidal zone.

Although the flat, gradually shelving beaches of Oceano, Pismo Beach, and Morro Bay are the natural center of distribution of this clam, it is found from Half-Moon Bay north of San Francisco to Socorro Island off southern Mexico. Nowadays one can sometimes more easily find the Pismo on suitable beaches along Baja California than in the depleted beds of California itself. The Pismo is found only on pure-sand beaches exposed to the open ocean. It not only endures surf, but absolutely requires it. Attempts to keep Pismos in convenient lagoons have always failed. The Pismo has so accustomed itself to the high oxygen content of the constantly agitated surf that it quickly dies when moved to quieter waters. Sometimes it shares these surf-beaten beaches with the Razor Clam, *Siliqua lucida*, and sometimes it is the only edible thing to be found on such beaches.

One of the attractions of the Pismo Clam is, as I have mentioned, its great size. A legal-sized one weighs one pound or more, and there are records of four-and-a-half-pound specimens. Of course, much of this weight is in the heavy shell, but there is still a sizable piece of edible clam in every Pismo. One would expect a clam living in pure sand and where the water is always roiled and sandy to be too full of sand to be edible, but not so. Examine a Pismo closely and you will see why. The short siphons are united except at the very end. The incurrent siphon at first glance appears to have no opening, but under even low magnification one sees that it is really covered by a living screen formed of branching papillae with openings so fine that they exclude the sand while allowing the water with its load of food and oxygen to pass.

The only thing anyone could have against the Pismo Clam is its increasing rarity and the subsequent strict regulation governing its digging. It is my hope that this book will open many eyes to the other wonderful seafood possibilities along the shore and remove some of the pressure from this really fine clam.

THERE REALLY IS A PURPLE CLAM
Sanguinolaria nuttalli

Unlike the famous "Purple Cow," the Purple Clam actually exists, and when properly de-sanded and prepared for the table, it is truly a noble dish. This is a Southern species, being found in suitable habitats from San Pedro to San Diego. It is found lying about a foot below the surface of tidal sand flats that are protected from high surf and swift tidal currents. It is a cleanly creature and doesn't like even small amounts of mud in the clean, water-bearing sand or sandy gravel that is its favorite home.

The Purple Clam is of good size, some specimens measuring as much as 4½ inches across, but most of those taken will be between 3 and 4 inches in length and oval in outline. They are not as rounded and thick as the cockles, but neither are they as thin as the *Macomas*. Like the Bent-Nose Clam's, the shell is slightly bent to one side at the siphonal end, but in the Purple Clam the bend is to the left, and the right valve is flatter than the left. This doesn't mean that the Purple Clam is merely a left-handed *Macoma* dressed up in bright colors, for there are many differences between the two species despite some superficial resemblances. The dried shells of these clams are easily recognized by a distinct purplish tinge both within and without, but, while the clam is alive, the outside of the shell is covered with a glossy, brown periostracum that hides its unique coloring and gives it a varnished appearance. The valves are thin and easily broken, especially near the edges. As is to be expected in such a deep-burrowing species, the siphons are long, but what is unexpected is that they are separate throughout their length, confounding those people who insist on calling a clam's siphons its "neck." The incurrent siphon is slightly longer than the excurrent one, and both look something like slender, white worms, an appearance that puts some people off when these clams are served whole with the siphons intact. Better chop those siphons up.

The Purple Clam sends its long siphons up through two separate holes that reach the surface from 1 to 3 inches apart, and the presence of these paired holes on the surface of sandy flats betrays the clam beds beneath.

The problem is to get the clam to the surface without breaking its thin, brittle shell. A friend once demonstrated his method of getting Purple Clams. We first located an area on a clean sand flat where dozens of twin holes showed there were clams underneath, then my friend shoveled off a 10-inch layer of the surface sand. As he was dressed in bathing trunks, he stepped into the wide, shallow hole he had made and began feeling through the soft, water-filled sand with both hands and feet, tossing out a clam every few moments. From about two square yards of flat we quickly took four dozen clams.

Since my friend had showed me his trick of treading out Purple Clams, I showed him my trick of de-sanding them. We put them in a hardware-cloth basket, such as I use for cleansing the mud and sand from *Macomas* (page 179), and hung them from a pier in clean seawater for two days. They were entirely free of grit, and the flavor was wonderful. Purple Clams taste much like other clams, but the good components of the flavor seem to be accented, and the poorer ones muted, making them as fine a clam as I have ever tasted. They would command good prices on the market, if only their shells were not so easily broken, and if they could close their valves tightly to retain their moisture. This inability to close their valves completely, while it makes them poor shippers and poor keepers, does make them very easy to open. A thin knife can easily be slid between the valves to sever the adductor muscles, and then they will open of themselves.

Open Purple Clams raw, chop the meat, and use it in stuffed clams or clam fritters. The flesh is tender and sweet, and should not be overcooked if you would enjoy it at its best.

A book with the limitations of the present volume cannot possibly give an exhaustive treatment of all West Coast mollusks. You will find information on other West Coast edible bivalves in the chapters on Boring Clams, Cockles, Razor Clams, Oysters, Scallops, and Wedge Shells, but all these are a small fraction of the total number that could be mentioned. There are more than five hundred species of bivalve mollusks on the West Coast of North America, north of Mexico. The few dozen species that I have described are those most commonly used as food, but there are others that would make possible food for man, and all of them are of intense interest to the naturalist, the shell collector, and those who get their joy merely from observing the strange forms that life takes in that living strip of shore between the tides.

20. The Epicurean Abalone
(*Haliotis* species)

ASK a covey of Californians to name their favorite seafood and most of them will answer, "Abalone steak." Even before the arrival of the white man, the Abalone was highly appreciated by the California Indians, and the beautiful iridescent shells were an easily available source of wealth, being traded far inland. When the Forty-Niners arrived they found Abalones in prodigal abundance; one could gather bushels of them at every low tide. The Chinese came and were delighted with these bountiful beds of Abalones, for this was a favorite seafood in China, but overfishing had made them rare along the Asiatic coast ages before. They gathered and dried the California Abalones by the ton to ship back to the Celestial Empire. The succeeding waves of people arriving from the Eastern States loved Abalone steak at the first taste, so fishing for them and preparing the steaks for market was soon a thriving industry.

The Abalone is a slow-growing creature, and couldn't stand these multiple attacks. All this unrestricted commercial fishing soon began to deplete the supply. The state was forced to pass drastic conservation laws to keep this valuable resource from being destroyed altogether. Now, there are closed seasons and severe limits on the size and number that can be taken, while drying, canning, or even shipping them out of the state is strictly prohibited. Today it is only

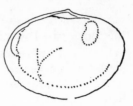

Purple Clam. *Left, exterior view.*
Right, interior of shell. Length:
about 3 to 4 inches.

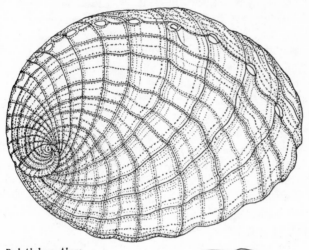

Red Abalone. *Above,*
exterior view. Below,
interior of shell. Size: to
about 7 inches in diameter.

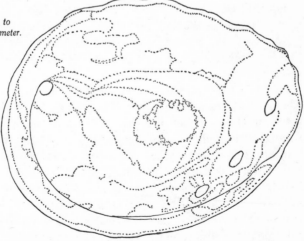

residents of the Pacific coast and tourists to this area who may enjoy this prize of all mollusks.

Even with consumption limited to local residents, Abalones of legal size are becoming increasingly hard to find. As soon as an Abalone reaches the legal size limit, a sportsman or commercial fisherman is there to snatch it from the rocks. The chances of finding one of legal size in the intertidal zone, or even in wading depth, are so slender as to be hardly worth the effort. It was these low intertidal areas and adjacent shallows that were once the very center of Abalone distribution, but now one must go farther out and dive for them to get the big ones. There are still many Abalones in the shallow areas, and even above the lowest tides, but practically all of them are undersized and thus protected by law.

The Red Abalone, *Haliotis rufescens*, the largest of California Abalones, is found from north of San Francisco to Lower California, reaching its finest development around Monterey. Most people are familiar with Abalone shells, for they are favorite decorations. Examine one of these shells closely and you will see that the Abalone is a member of the snail clan. The coiled spire is greatly flattened, and the entire bottom of the shell is open, but the Abalone is obviously a gastropod, and Abalone steaks form a notable exception to the usual American prejudice against eating snail-like creatures. Near the left margin of the shell there is a row of holes, the number differing with the various species. On the Red Abalone the number is 3 or 4. Under these breathing holes is a respiratory chamber containing 2 gills. The Abalone moves about and holds fast to the rocks by the powerful muscles of its large foot. It is this fleshy foot that is sliced into the delicious Abalone steaks.

Because of its commercial importance, the life history of the Red Abalone has been studied. It takes six years for a female to reach spawning age, and then it is only 4 inches in diameter. At this age a female will produce about 100,000 eggs per season, but by the time they are ten to twelve years old, and 7 inches in diameter, each one will produce upward of two million ova between the middle of February and the first of April, which is their breeding season. The ova and sperm are merely released in the ocean, and fertilization occurs by chance encounters in the water. If left undisturbed, the Red Abalone grows to 9 inches in longest diameter, but on the overfished California coast it is an extremely lucky or very well-hidden Abalone that ever reaches this size. The life cycles of other species of Abalone are thought to be similar to those of the Red.

From Santa Barbara south, and along the Pacific side of Lower California, the Green Abalone, *Haliotis fulgens*, is found. In spite of its common name, the Green Abalone is a dull reddish brown on the outside, but it can be distinguished from the Red Abalone by its 5 to 6 (usually 6) breathing holes, all elevated like little volcanoes. It is a bit smaller than its red cousin and, while the Red Abalone is usually covered with growths of various kinds, the Greens keep their shells clean. It is quite as good as the Red Abalone.

Occurring with either of the above Abalones, one may find the Black Abalone, *Haliotis cracherodii*, but usually Blacks prefer more surf than the other two. Their favorite habitat seems to be vertical fissures in surf-pounded rocks. They don't grow quite as large as the Reds, but the legal minimum for collecting them is 5½ inches. There are 5 to 8 breathing holes, and these are not elevated as are those on the Greens. The shell of the Black is clean and shining, a greenish black on the outside and silvery with red and green reflections on the inside. The steaks are slightly tougher than those of other species, but they taste just as good if pounded a little more thoroughly before cooking.

Far to the north, along the outer coast of British Columbia and Alaska, is found the prettiest Abalone of them all, the Pink Abalone, *Haliotis kamtschatkana*, only about 4½ inches in diameter at maturity, with a wavy pink shell. This little Abalone seems to be rather particular about its environment. It does not occur up the Inland Passage, but prefers protected or semiprotected waters on the outer coast. When located, it is often found in abundance, for the Pink is very gregarious. Its preference for protected waters makes gathering it less strenuous and hazardous than the hunting of most Abalones. The lucky yachtsman who gets the opportunity to cruise the beautiful and almost unspoiled coast where this species is found has a treat in store for him, for this is one of the most delicious of all Abalones, so tender that it requires little if any pounding to make it ready for the pan.

The mere mention of the word "Abalone" brings pictures of foggy California dawns when, shivering, I launched a paddle board and hand-propelled it out to a sunken reef, looking for Red Abalones. The sea is usually calm at this time, and one can safely poke about rocks and ledges that will be covered with breakers after the breeze kicks up a surf. The best Abalones are often found on the under sides of ledges where they are not visible from the water's surface. I had a glass-bottom viewing port in my paddle board and when I found myself over a likely spot I would adjust my diving mask, slip overboard, and investigate. The Red Abalone is hard to see underwater, for it invariably carries a small forest of hydroids, algae, and other marine growth on its shell. When you see one of doubtful size, make sure that it is of legal length before prying it from the rocks. I carried a flattened steel bar about 20 inches long with a plainly filed mark on it just 7 inches from one end. The water has a magnifying effect and an Abalone that looks as though it's 10 inches across under water may turn out to be just under the legal limit if brought to the surface before it is measured. When you find a legal-sized one, slip your bar quickly under it and pry it from the rocks before it has a chance to get a firm grip. The stories about Abalones catching people by the fingers or toes and holding them for hours until the tide has come in and drowned them are not very well authenticated, but an Abalone wouldn't have to hold on long for the results to be fatal if you mask-dive without an air supply, so if you do, keep your fingers out from under that shell. On the rare occasions when I have found a legal Abalone abovewater, during extremely low spring tides, I have sometimes slipped my fingers under the shell and jerked the Abalone suddenly from its moorings, but I wouldn't try that while diving. After prying an Abalone from the rocks, I would lay it flat on the paddle board. It would then seize on the smooth surface of the board, as it does on a rock, and not try to escape, apparently not realizing that it was being "taken for a ride."

To clean an Abalone, remove it from the shell and cut away the viscera. I recommend that many shellfish be eaten entire, but not the Abalone. Being a strict vegetarian, the Abalone has a very large green gut that is anything but edible. Slice the huge foot, which is the edible part, into steaks about ⅜ inch thick. Some Abalone fanciers place these steaks on a firm surface and pound them with a rolling pin or meat hammer. Others put them between two pieces of clean, strong muslin and pound them with a wooden mallet. However it is done, this pounding should be thorough, for this is what makes the difference between a tough, barely edible Abalone steak and one that is among the finest culinary creations that ever came to the table.

After pounding, dip the steaks in egg that has been beaten with a little water, dredge in seasoned flour or bread crumbs, and fry to a light brown. It is said that Abalone also makes good chowder, but I cringe at the thought. It would be too much like making an ordinary beef stew of *filet mignon*. Make your chowders of clams, mussels, and other less kingly creatures, but reserve the lordly Abalone for those tender and delicious Abalone steaks. And, unlike most seafood, Abalone steak is actually improved by being kept in the refrigerator for a day or two before being cooked.

21. The Edible Limpets

THE knowledgeable seaside camper can enjoy a wonderful variety of delicate and wholesome foods unknown to those whose seafoods are limited to the kinds that commonly appear in the fish markets. Some of the finest seafoods in the world are never gathered commercially, for a number of reasons. The seaside gourmet who has developed an epicurean appreciation of these rare delicacies may indulge in gastronomic delights unknown to ordinary mortals. A good illustration of these little-known but perfectly delicious seafoods is the Owl Limpet.

Several limpets are eaten in various parts of the world, but the finest of them is undoubtedly the huge Owl Limpet, *Lottia gigantea*, of the Pacific coast from northern California to southern Mexico. The largest and best ones are found on surf-swept rocks in the intertidal zone. This is a giant of the limpet tribe, often attaining a diameter of more than 3 inches. Like all its kinsmen, the Owl Limpet has a single conical shell and depends on the rock that is its home to close up the bottom of its portable house, or maybe I should say "tent," for a limpet shell looks like nothing so much as a tiny, squat, Indian tepee. However, the limpet doesn't pitch its tent on a new spot every night, but lives at one specific place on the rock where it has worn a scar that fits its own shell and and no other. Limpets may crawl about and socialize while using their file-like tongues to rasp the algae that form their food from the rocks, but naturalists who have marked large numbers of limpets and their resting places find that each limpet always returns to its own scar, although, in a few square yards of rock there may be thousands of scars that appear identical to us.

In rocky situations in California and Lower California, Owl Limpets are commonly found so abundant that they could be gathered in large quantities if one had any legitimate use for so many. To remove a limpet from its rocky home, suddenly slip a thin-bladed knife or a sharpened putty knife under the shell and pry the limpet loose. If you touch the shell first, or otherwise give the limpet warning, it will pull itself so tightly against the rock that it will be impossible to remove it without breaking the shell. It has been estimated that a limpet, in holding onto its own rock scar with its flat foot, can exert a pull of 70 pounds per square inch of surface, and for a large Owl Limpet this would add up to a total pull of more than 400 pounds.

Owl Limpets are seldom eaten in California, but they are highly esteemed by the Mexicans of Lower California, who know a good thing when they taste it. Each Owl Limpet will yield just one small steak about 2 inches across, but when this is properly prepared it is delicious, being of finer grain and more delicate flavor than even the highly prized Abalone. Each little Limpet Steak should be placed inside a fold of muslin and pounded on a wooden surface with a mallet or rolling pin to make it tender, then dipped in egg that has been beaten with a dash of water, dredged in fine bread crumbs, and fried to a golden brown. When cooked perfectly, these limpet steaks can be cut with a fork and have a rich but delicate flavor that is just right.

The next time you are in Southern California and a native son starts bragging about the wonders of Abalone steak, practice a bit of one-upmanship by finding some Owl Limpets and treating him to a taste thrill that makes Abalone seem ordinary. Unless he is a student of marine life, he probably doesn't even know this fine food is available.

Besides the Giant Owl Limpet, the Pacific coast has a rich assortment of smaller limpets, many of them found all the way from Mexico to the Bering Sea. All of these are in the genus *Acmaea* and there are many species, ranging in size from minute to more than 2 inches across. The various species differ from one another in proportionate height, coloring, roughness and sculpturing of shell, and other details, but they are so closely related that the species tend to grade into one another.

These *Acmaea* species are usually dismissed as too small for food, but this is nonsense. In Hawaii, a related *Acmaea*, smaller than several of the West Coast species, is so highly esteemed that the Hawaiians will expend great effort and no little risk in gathering them from the surf-beaten rocks where they make their home. Old Hawaiians, who remember when a luau, or native feast, was something besides a tourist attraction, consider a luau without *opihis*, as they call this little limpet, distinctly second-rate. While living in Hawaii I learned to enjoy this little conical shellfish so well that the last vestige of any prejudice I may ever have had against eating limpets disappeared completely.

My favorite way of cooking *opihis* was to fill an edged cookie sheet with ordinary table salt and bed the limpets in it with the open side up. The salt wasn't for seasoning, and never came in contact with the meat; it was merely a means of getting the little conical shells to sit upright with a level rim. I then put a few drops of Melted-Butter Sauce, strongly laced with Tabasco, on each *opihi* with a medicine dropper and put the pan under a hot broiler for 10 minutes. The pan was brought to the table while the juice in each little inverted cone was still boiling hot. I have not had an opportunity to try this method on West Coast limpets since returning from the Islands, but I'll wager it will work as well with several of the California species as it does in Hawaii.

The common limpet of the shores of the British Isles, *Patella vulgata*, is little if any larger than some of the *Acmaeas* of our own West Coast, and it is—or at least *was*—eaten and relished. They were collected by tapping a square-pointed blade under the shell with a wooden mallet, and one writer records that, in some stretches of the English coast, low tides were characterized by the constant sound of tapping made by the limpet gatherers. Old English cookbooks include recipes for preparing limpets, and one of these is so unusual that I must pass it on to you, although I must admit that I have not tried it. The limpets are placed in a large baking pan, this time open side down, and allowed to seize on the bottom of the pan as they do on their native rocks. Dry hay is piled over them, and about 20 minutes before they are to be eaten, the hay is set

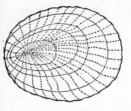
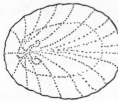

Owl Limpet. *Left, upper side; center, under side; right, side view. Size: about 3 inches in diameter.*

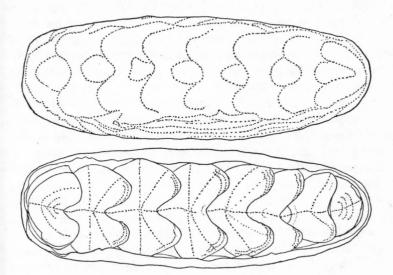

Above, Giant Sea Cradle. *Upper and under sides. Length: to 13 inches.*

Below, Magdalena Chiton. *Upper and under sides. Length: 2½ to 3 inches.*

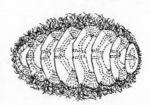

afire. It quickly burns down, leaving glowing embers about the shells, and it is claimed that this furnishes enough heat to cook the limpets exactly right.

To make a Limpet Chowder, put a quart or more of limpets into a covered kettle with 1 cup of water, and steam for 10 minutes. Remove the limpets and save the broth. As soon as the limpets are cool enough to handle, remove the meat and discard the shells and viscera. Grind the meat in a food chopper with a medium blade, then proceed exactly as in making Quahog Chowder (page 21).

Another way to prepare limpets is to remove the edible meat from raw limpets and grind it fairly fine in a food chopper. Stir and work this meat with the fingers until it assumes a spongy texture, then form into little balls or cakes with dampened hands. These little cakes are very good when fried gently in butter and served with green onions, or, you can mix chopped green onions and chopped green peppers with the meat before you make it into cakes. These Limpet Balls are also very good dropped back into the broth in the steaming kettle and boiled for 10 minutes, then served as a soup.

Compared to the Pacific coast, our Eastern shores are very poor in limpets, there being but one species large enough to interest us. This is *Acmaea testudinalis*, commonly called the Atlantic Plate Limpet or the Tortoise-Shell Limpet, and even this one seldom attains a diameter of more than 1½ inches. I have collected these limpets in Maine, but I never found enough at one time to try cooking them. I have eaten them raw and they seem edible enough, but there is only a very tiny piece of meat, and it is not as sweet as the Hawaiian *opihi*.

Even along the Pacific shores where these creatures are plentiful, limpets will never make a major contribution to the cuisine of the coast dweller, but to the seaside camper with an adventurous palate they can furnish an interesting and delicious change of menu.

22. Chitons or Sea Cradles

FEW people have ever thought of the chiton as an edible mollusk, but peoples as far separated as the American Indians of the Pacific Northwest and the natives of the West Indies have independently discovered that they were good food. Although most of our East Coast chitons are too small, or too difficult to gather in quantity, to furnish much food, none of them are poisonous, and my own sampling in this area indicates that there is nothing amiss with the taste buds of those who have enjoyed chitons.

As with several other seafoods, the people along the Pacific coast have the advantage over Easterners. On the North Atlantic coast we must be content to sample little chitons ranging in size from minute to 2 inches long, but on the West Coast there are many species larger than that, and some attain an enormous size, for a chiton. Among these is the Giant Sea Cradle, or Gum Boot, *Amicula stelleri*, formerly called *Cryptochiton stelleri*, the largest chiton in the world, attaining a length of 13 inches. These and several other large species were eagerly sought by the West Coast Indians, and they became favorite seafoods of the Russians who first settled southeastern Alaska. The Giant Sea Cradle is found from Alaska southward to Southern California, and westward to Japan.

Chitons are usually found on the under side of rocks at about the low-tide line, or in tide pools. At night they come out and feed on the algae covering the rocks, for they are strict vegetarians. The different species vary widely in size and color, but they are much alike in shape, structure, and general appearance. Oval in outline, the mollusk consists of 8 overlapping plates surrounded by a flexible girdle, allowing the creature to curl up like an armadillo when it is dislodged from its footing. Underneath the shell there is a long, flat, fleshy foot, and the internal organs are between the foot and the shell. When a chiton dies, the plates forming its shell come apart and often are found on the beaches. Children who pick up these shells usually call the middle ones Butterfly Shells, and the two rounded end ones, which look something like upper dental plates, False-Teeth Shells.

The Giant Sea Cradle differs from most chitons in that its girdle covers the shell plates, and at first glance it appears shell-less. This big chiton, often a foot long, is a dull brick-red in color. In the spring I have seen great numbers of this giant species congregated just at low-water line on rocky beaches of northern California. One could have collected dozens of them in a few minutes. Other times I have searched in vain for them in the same areas where I had found them before. Most chitons are light-sensitive and are much more active at night than during the day, but this giant form seems to be bothered by light less than most species and will often be found out on the rocks or in the tide pools on a foggy day. On bright days, seek them by turning over loose rocks at extremely low tide.

The edible part of the chiton, as with Abalones, Owl Limpets, and other "open-bottomed" mollusks, is the fleshy foot. The foot of a giant specimen is large enough to furnish a steak. Turn the chiton on its back and run a sharp knife around between the shell and the flesh. Discard the shell and the internal organs. You will be disappointed in the yield, which is only one small, thin steak per chiton. The Northwest Indians solved the preparation problem by cutting the meat into bite sizes and eating it raw, often right on the spot within a few minutes of the time it was caught. They had the right idea, for chiton steaks, to be good, must be fresher than fresh. If you are collecting chitons to carry home or into camp, carry a portable ice chest and some crushed ice to the beach. Pack the creatures into it as fast as they are caught, then clean and eat them as soon after as possible. If there is to be any time between cleaning and cooking, put the steaks on ice until you are ready to pound them and drop them into the pan. If chiton steaks are allowed to stay in a warm place, they develop a strong fishy odor that repels most people.

The Indian custom of eating chiton meat raw should not be disgusting to those of us who are accustomed to eating raw oysters and clams, viscera and all. Thoroughly chilled chiton meat, cut into bite sizes and dressed with lemon juice or cocktail sauce, is quite palatable to most tastes. If you prefer cooked seafood, put the steak in a fold of muslin, lay it on a chopping board, and pound it vigorously with a wooden mallet or the rolling pin. Fold it over, or better yet, place two steaks together, and pound again. Beat an egg with 2 tablespoons cold water, dip each doubled steak in the egg, then cover with fine bread crumbs and fry to a golden brown on both sides. Serve while still sizzling hot.

No other species of chiton besides *Cryptochiton stelleri* can furnish a piece of meat large enough to be called a steak, but there are other Pacific species that would be considered enormous anywhere else in the world. The Magdalena Chiton, *Ischnochiton magdalenensis*, and the Black Chiton, *Katharina tunicata*, are both found from Alaska to Catalina, and both reach a length of 2½ to 3 inches. The Conspicuous Chiton, *Ischnochiton conspicuus*, plentiful from Monterey to the Gulf of California, grows even larger, reaching a length of 4 inches. The Black Chiton was eaten raw by the Northwest Coast Indians and it makes even better chowder or soup than the Giant Chiton. I have not tried the other two species, but some brave soul on the West Coast should test their culinary worth.

The natives of the West Indies gather and eat several smaller species of chiton. There chiton meat is known as "sea beef" and it is fried, boiled, or used in soups and chowders.

The way to handle all these smaller species of chiton is to put about 2 quarts of washed chitons into a covered kettle with 1 cup of water and steam for 10 minutes after the water starts boiling. Strain and save the broth. When the chitons are cool enough to handle, remove the meat and discard the shells. The yield of meat is small, which is probably one reason that chitons have never become a popular food among us, but what there is of it is very good. To make Chiton Soup, melt 2 tablespoons butter in a heavy frying pan or kettle, then add the cleaned chiton meats, 1 onion, 1 potato, 1 carrot, 1 clove garlic, ¼ cup green pepper and ¼ cup celery, the meats and all vegetables chopped fine. Cover and sauté until the onion is clear and yellow, but keep the heat low so none of it browns. Add enough water to the broth in the steaming-kettle to make 1 quart, and add this to the vegetables. Season with salt and freshly ground black pepper. Simmer for 30 minutes, or until the carrot and potatoes are soft, then add ½ cup white wine. Dish into small, deep bowls, and sprinkle with minced parsley. Float a round of buttered toast on each serving, and place it immediately before your guests.

23. Hunting the Wild Goose Barnacle

(*Lepas* and *Mitella* species)

THE Goose Barnacle is one of the most widely known of all intertidal creatures, for it is worldwide in distribution. An accomplished transoceanic hitchhiker, it steals rides on floating logs, timbers, and the bottoms of man-made ships, and it is now established
on every shore where conditions are suitable to it. Despite being
almost universally known since prehistoric times, the Goose Barnacle
has been one of the most misunderstood creatures that ever lived.
Its strange, hybrid appearance deceived even the most learned men
of old. They saw a chalky-white, calcareous, two-valved shell, each
valve made up of several joined plates, all mounted atop a firmly
anchored, fleshy stalk 1 to 4 inches high. Those who bothered to
open the shell were only further confused, for they saw an almost
formless creature that seemed to be partly covered with slender,
flexible, featherlike fronds. Through the ages, this puzzling life-form
was variously classified as a plant, a tree bearing strange fruit, a
mollusk related to the clams and mussels, and as the young of certain wildfowl that had never been observed to nest or lay eggs. Of
course it was none of these.

I imagine it was the Italians and Spaniards who first suspected the true relationships of the Goose Barnacle, for they judged, not by appearance or habits, but by taste. Other Europeans did not consider the Goose Barnacle edible and therefore did not have access to this source of information. This aberrant creature doesn't look or act at all like other crustaceans, but the pink meat in the fleshy stalk tastes too much like lobster, crab, and shrimp not to be related to them. It was not until 1830 that a British scientist, J. Vaughan Thompson, finally demonstrated that the Goose Barnacle develops from a typically crustacean larva, and this creature's rank and place among his kind were settled once and for all. As Thomas Huxley once so aptly said, "A barnacle may be said to be a Crustacean, fixed by its head and kicking food into its mouth with its legs."

It was the most ridiculous of all these ancient theories about the nature of the Goose Barnacle, namely that it was the young of certain wildfowl, that had the greatest circulation and was most widely believed. Indeed, the term "barnacle" was at first the common name of a small Arctic goose, *Branta leucopsis*, which is a winter visitor to Britain. As this goose nests in the remote Arctic, its eggs had never been observed, so this absurd myth helped to explain both an eggless goose and that other unexplainable life-form that clustered on driftwood and ships' bottoms; so it was that the term "barnacle" gradually began to be applied both to the bird and to its supposed young.

Amazingly, we have accounts of the impossible transformation of a barnacle into a bird. In the sixteenth century, a Dr. Priest translated a Flemish work on natural history into English, but died before he could publish it. This translation fell into the hands of a lovable liar and plagiarist by the name of John Gerard. He altered the sequence of the items, added a few of his own observations, and had the book published as his own work. As a suitable ending to his massive volume, he writes about "this woonder of England" the "Barnakle Tree," or "the Breede of Barnakles." He tells of reports that there are trees in the north of Scotland and in the "Orchades" (wherever they are) that bear shells that open and hatch out barnacle geese. Apparently acting on the theory that pictures cannot lie, he illustrated a section of his "Historie" with a woodcut of a "Barnakle Tree," showing strangely branched barnacles hatching into geese. This nonsense continued to be repeated, republished, and believed for almost two centuries.

The Goose Barnacle is almost never eaten in America except by a few Italian and Spanish immigrants. I have heard that they are sometimes offered on the San Francisco market, but as John Gerard would have said, "I cannot absolutely avouch" for this, for I have never seen them there. In 1916 the California State Game and Fish Commission recommended the Goose Barnacle as a food animal and published a recipe for preparing it. It was on coming across this old recipe that I first went in search of the wild Goose Barnacle. The kind I gathered was the Pacific Goose Barnacle, *Mitella polymerus*, which crowds the rocky shores of the West Coast by millions from Alaska to Lower California. I believe the various species of *Lepas*, common along the East Coast from Nova Scotia to Florida, would work as well, for they are very similar in appearance.

I cut through the stalks as close to their anchorage as possible without getting dirt or grit, washed them thoroughly, put them in a covered pan with a little water, and steamed them for 20 minutes. Then I removed the shells, pulled out the feathery legs, and peeled away the tough, rindlike skin. There was much more waste than edible meat, but since this is true of all other crustaceans, I didn't let it worry me. The meat that was left tasted a little like lobster, but I must admit that it wasn't as good. On my first try at serving it, I diced a cup of the Goose Barnacle meat and mixed it with 2 tablespoons minced onion, 1 teaspoon white wine vinegar, ⅛ cup mayonnaise, and ¼ teaspoon salt. I then split and buttered 2 hamburger buns and spread them thickly with this mixture. Each of these Mock Lobster Rolls was wrapped in its own piece of foil, and they were baked in a 350° oven for 20 minutes. Not bad at all; in fact it was an excellent luncheon dish.

The next time I tried Goose Barnacles, I steamed and cleaned them the same way, then chopped the meat and made it into a Goose Barnacle Newburg, using the recipe on page 108, which is heavily laced with sherry. It was excellent when served over hot buttered toast. I will not maintain that Goose Barnacle will be the best seafood that you ever tasted, but it is good meat with a definite crustacean flavor, and where abundant could easily become the poor man's Lobster.

24. Grunions, Blennies, and Lancelets:

HOW TO DIG A FISH

DURING the second, third and fourth nights after the full moon, in the months of March, April, May, and June, the Grunions spawn. The Grunion, *Leuresthes tenuis*, is a small, smeltlike fish about 6 inches long. Many of the inhabitants of the upper beach, such as the Rough Periwinkle and the dune crab, live their lives on shore but go back to the sea to spawn, but the Grunion has reversed the process; it lives in the sea and comes ashore to spawn. This spawning is as perfectly timed as the movements of those celestial bodies that cause the tides. On only the nights mentioned above, and just after the high tide has turned and is beginning to ebb, the Grunions start coming ashore. They come in pairs, male and female, riding the very crest of the wave. When the wave flattens out on the beach, the Grunions actually swim ahead of it, sliding along on the sand with only their tails in the following wave. When the water reaches its highest point on the shore and starts to recede, the female digs her tail into the sand and the male throws himself across her in an arched position. Here they hold fast while the wave runs back down the beach. For the time between two waves they are in an alien element, sometimes many feet from the nearest water. During this brief period the female deposits her

eggs beneath the sand and the male fertilizes them. When the succeeding wave laps toward them, they slip into it and escape back
into the sea.

The eggs incubate for two weeks in the warm sand of the upper
beach. Then comes the even higher new-moon tide that washes
these eggs from the sand, when they immediately hatch and the
microscopic young swim out to sea. They are fast-growing and early-
maturing fish, and by next spring these newly-hatched young will
have grown into adults and will be back to take part in the spawning
ceremony.

It is an amazing performance. If the Grunion came in on a rising
tide, or even one wave too early, succeeding waves would wash the
eggs from the sand and carry them back into the water where they
would soon be devoured by the tiny, planktonic monsters that swarm
everywhere in the sea. On the other hand, if they spawned during
the higher new-moon tides, the eggs would lie in the sand for four
weeks before being released into the water, and by this time the
embryos would all be dried out and dead. Their instinctive timing
device leads them unerringly to shore at the only times that this
kind of spawning is possible.

The Grunion is found on the Pacific coast from Monterey Bay to
far down in Lower California. To Southern Californians who love
the seaside and appreciate good food, Grunion spawning is a time
of high holiday. They go to the beach by the hundreds to catch and
eat these tasty little fish. It is a strange scene lighted by the bright
moon and many beach fires. Sometimes hundreds of pairs of Grunions will come in on a single wave, and then there is a mad scramble
to capture them before the following wave releases them back into
the sea. Dip nets, saucepans, hats, and hands all become fishing
equipment. Most people just run along the wave edge with a pail,
flipping the slippery fish into it with their bare hands.

Many maintain that the Grunion needs no cleaning whatever.
Maybe I'm being over-nice but I must admit that I prefer Grunions
with the heads and entrails removed. This is not a hard job, and the
right tool to use is a pair of sharp kitchen shears. Snip off the head
and tail. Run your fingers down the outside of the fish and press the
entrails out where the head was before you cut it off. I have cut
Grunions open after cleaning by this method, and I can vouch for

the efficacy and completeness of it. Even the intestine that runs back to the vent breaks cleanly and comes out entire.

Most Grunions are cooked right on the beach where they are caught, and this is definitely part of the fun of Grunion time. Spit the fish on a sharpened stick and roast over coals just as you roast hot dogs. Or drop the cleaned fish into a frying pan with a little hot fat on the bottom of it and fry until it is delicately browned all over. When it is cooked in this manner, the bones are not at all noticeable, and the fish can be eaten end-on like a stick of candy. These little fish cooked over a campfire a few minutes after they are caught are exceedingly delicious. Some like to add a light sprinkling of salt, but others find their own sea-saltiness sufficient.

If you capture more Grunions than can be consumed at the beach, there is no law against serving them at home. Take your portable ice chest to the beach, clean all excess Grunions as soon as they are caught, and pack them in crushed ice immediately. When handled this way they will be almost as good on your dining room table as they were at the beach. French Fried Grunions will be appreciated by anyone who likes seafood fried. Wipe each fish with a cold, damp cloth and roll it in fine bread crumbs. Dip the fish in egg that has been beaten with 2 tablespoons cold water, then dip into the bread crumbs again. Drop the crumbed fish into deep fat that has been heated to exactly 375° and cook for just 3 minutes. Drain on paper towels, and serve hot with lemon wedges and parsley.

One of my favorite recipes for preparing this little fish is Pickled Grunions. Wipe the fish clean with a damp cloth, place in a colander, and steam over boiling water for just 3 minutes. Arrange the steamed fish in a glass casserole that has a cover, and pour over them a sauce made by boiling together for 5 minutes 4 cups vinegar, 1 choppd onion, 2 bay leaves, and 1 tablespoon mixed pickling spices. Be sure the fish are covered with the sauce, even if it means making more sauce, using the same ratio of proportions, and pour on the sauce boiling hot. Cover, but do not seal, and leave it in the refrigerator for three days before eating. These make excellent appetizers, hors d'oeuvres, or cocktail snacks, or they can be served as the fish course of an elaborate meal.

Another fish that can be caught in the intertidal zone with no equipment is the Blenny. The little fish called Blennies are of several

Above, Conspicuous Chiton. *Upper and under sides. Length: to 4 inches.*
Below, Black Chiton. *Upper and under sides. Length: 2½ to 3 inches.*

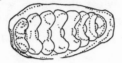

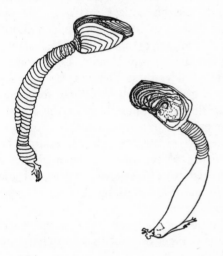

Goose Barnacle. *At right, view of under side. Size of shell, about 1¼ inch long.*

Lancelet. *Length: about 1½ to 3 inches.*

Smelt. *Length: about 8 inches.*

genera and numerous species and sub-species, so instead of going into a long dissertation on scientific names and recognition keys, we will merely say that any little eel-like fish from 3 to 10 inches long that can be captured in the intertidal zone is a Blenny. In color they range from light olive to black; some species have bright stripes of white or yellow, while others have mottling of lighter colors near the tail, and still others are solid-colored. One or more species are found on rocky coasts from Alaska to the tip of Lower California.

I first encountered Blennies while gathering mussels on a rocky shore of Puget Sound. I pulled a cluster of mussels from a rock and was startled by a half-dozen little greenish-black eel-like fish flapping noisily and excitedly in an attempt to take cover under another cluster of mussels. I captured one and examined it. It was about 6 inches long, laterally compressed and very slender. On inquiry, I discovered that many people consider these Blennies very fine fare, so I returned to the beach and gathered enough of them for a meal. At first I continued pulling clusters of mussels from the rocks in search of them, then I happened to turn over a rock and under it were four good-sized Blennies in sand that was barely damp. I had never before had the experience of catching fish from almost-dry land, and became so fascinated with this new sport that I took more Blennies than I actually needed, something I seldom do when helping myself to edible seashore life.

My first experiment with cooking Blennies was disappointing. I cleaned and fried them in the manner described for Grunions, and, while they didn't taste bad, the smell while they were cooking was too reminiscent of the odor of a seaweed-covered beach at ebb tide to be really appetizing. I still had plenty of Blennies, so next I cleaned about a pound of them, laid them in a deep glass dish and covered them with a soaking mixture made of 1 cup cold water, ¼ cup vinegar, ½ cup salt, 1 teaspoon black pepper, and ⅛ teaspoon mace. I didn't arrive at this mixture by any kind of magic intuition, for it was a mixture that I often use on fish of many kinds when I want to improve or alter the flavor. After allowing them to soak for an hour, I rinsed the Blennies under cold water, wiped them dry, then dipped them first in egg, then in cracker crumbs, and fried them. These Fried Blennies were delicious. I have since cooked flavored Blennies without a coating, just dropping them one by one into hot fat, 375°, and French frying them for only about 3 minutes. These also were very good, and they had rather a pleasant odor as they cooked. When exploring the food possibilities of the West Coast, don't overlook these semiamphibious fish just because they are small.

The last of the intertidal fish we shall consider here is not really a fish—in fact, it is hardly a vertebrate. This is the Lancelet, or *Amphioxus*, a famous creature that is seldom seen. This tiny pre-fish, or almost-but-not-quite-a fish, has assumed a vast importance in biology, as it is considered an evolutionary link between the invertebrates and the vertebrates. Some such creature must have been ancestral to all vertebrates, and that includes you and me. This is enough reason to approach the Lancelet with respect, despite its insignificant appearance.

There are three common species of Lancelet on our shores, two on the Atlantic and one on the Pacific, but they are almost identical in appearance and habits, so we will consider them together. The Western Lancelet is *Branchiostoma californiense* and it is found from Monterey, California, to the Gulf of California in Mexico, but doesn't appear in the intertidal zone north of San Pedro. On the East Coast; the two very similar species are *B. virginiae* and *B. floridae*. With those specific names there is scarcely any need for me to tell you where they will be found.

In general appearance the Lancelet looks like a very slender, almost eel-like, tiny fish, from 1½ to 3 inches long as an adult. There are however many points of difference between the Lancelet and the true fishes. While it has membranous dorsal, caudal, and anal fins, there are no paired fins as in fishes. It is colorless, almost transparent, with a herringbone pattern when viewed from the side. This dimly-seen pattern is really its muscle segmentation seen through the naked, transparent sides. Its eyes are rudimentary and its mouth a mere opening, not equipped with jaws but surrounded with a fringe of hairlike, fleshy organs called *cirri*, which sweep the microscopic food on which it subsists into the mouth. A good idea of its general appearance can be gained from the illustration on page ooo, but un-fortunately a line drawing cannot show motion, and I have found that motion—very rapid, elusive motion—is one of the chief characteristics of the Lancelet.

My only experience with the Lancelet as food has been in California. I have always been fascinated by the strange food products that can be gathered from the sea and shore, and I avidly read every reference to the seafoods of other lands that I can lay my hands on. I read that the Chinese dredge up by the ton from shallow coastal waters Lancelets of a species very similar to our own. These bring premium prices on the Chinese market as a great delicacy. I have learned to trust Chinese taste, for I have seldom eaten a well-cooked Chinese dish that I disliked.

One day, while exploring the food possibilities of a sandbar in a bay near San Diego, I patted the sand smooth with my shovel as one does when seeking Razor Clams. Imagine my surprise when three little Lancelets popped out of the sand, then promptly dived back into it. Having my attention thus directed toward Lancelets, I began turning up shovelfuls of sand and really watching for them. Almost every shovelful had from 1 to 6 Lancelets in it, but that doesn't mean I caught them all. Each one would appear for only a split second, then promptly dive back into the sand. I have never seen a creature that could disappear so quickly. I finally managed to catch half a dozen, while several times that number escaped me. I soon saw that this hand method was not practical.

Next day I knocked the bottom from a wooden box and replaced it with coarse screen wire, then nailed some legs on the box so it would sit upright with the top about waist high. This time I took a companion along and we set the box in the water next to the sandbar. We had timed our arrival to coincide with low tide, so I was able to find the identical place where I caught the Lancelets the day before. I shoveled sand into the screen-bottomed box while my companion poured pails of water through it to wash away the sand. After several dozen shovelfuls, we would stop and pick out the Lancelets that were trapped on the screen bottom. There were always more sandworms than Lancelets but we would transfer the Lancelets to another container and wash the sandworms from the box before proceeding. In half an hour we collected a pint of Lancelets, which was plenty for the experiment we projected.

Since we had first heard of Lancelets as a Chinese food, we decided to concoct Lancelets Chinese-Style. We prepared a pot of fluffy oriental rice. I put a little peanut oil in the bottom of a heavy kettle and when it was hot I sliced into it a clove of garlic and a "thumb" of fresh, green gingerroot. When these began to brown, they were removed and discarded; we only wanted their flavor. Then the Lancelets were dumped into the hot oil and ¼ cup of soya sauce added. Lancelets need no cleaning whatever beyond washing off any clinging sand. The kettle was covered and shaken gently occasionally to stir the Lancelets. In a few minutes they lost their clear transparency, becoming white and opaque, looking something like tiny skinned fish. In 10 minutes we judged them done. We made beds of fluffy rice on the plates, then garnished them with the cooked Lancelets, pouring the juice over the rice as a sauce. We ate this meal with chopsticks, drinking many small cups of green Chinese tea, and ended the meal with fortune cookies and candied kumquats, and agreed that we had seldom dined better.

I do not know whether or not the Virginia and Florida Lancelets can be caught in sufficient quantities to use them for food, but I would not be surprised to find this possible. Such an inconspicuous creature is easily overlooked, and one who observes closely may find many more of them than is usually reported. The Eastern species are usually taken from the sand bottoms of shallow bays, from Chesapeake Bay to Florida, and it might be that an apparatus such as I used in California would yield a mess of Lancelets in Virginia. Let's try it sometime, shall we?

25. Fishing for Food
and Fun

I HAVE no desire to be a really stylish saltwater fisherman with expensive equipment and trophies on my walls. Nevertheless, I like to eat fish, and I have found that many of the neglected little fish, when skillfully prepared, are much better eating than the trophy-sized rarities sought by the Simon-pure sportsman. Then too, when I go fishing, I like to catch fish, consistently and continuously. I am not satisfied to spend time trolling or casting in order to catch a single fish, no matter how rare or large that one may be. I'll take the big ones when they come my way and thrill to them as much as any sports fisherman, but between the bragging-size fish I like to catch dozens of the smaller, sweeter kinds that will fit flavorfully into my frying pan.

I have a love affair with the shore and all its myriads of life-forms, edible and inedible, and the fish that interest me most are those that swim close to shore. I have caught far more fish from wharves, jetties, rocky points, sea cliffs, the banks of tidal streams and inlets, and from old bridges than I ever caught from boats or by orthodox surf casting, and I have spent very little money for equipment or commercial baits.

Recently, as a very pleasant part of the research for this book,

my wife and I took a seaside camping trip that extended from
North Carolina to New Brunswick and back. To prove my point
that expensive gear is unnecessary to the enjoyment of a great deal
of fishing fun, I took no fishing equipment with me. We stopped
at the first sporting-goods store and bought two completely rigged
hand lines, thirty-five cents each, two ordinary bamboo poles, a spool
of 12-pound test, monofilament line, a package of assorted small
hooks, two plastic bobbers, and a package of split-shot sinkers. The
whole bill came to less than five dollars.

The bamboo poles were about 15 feet long. I fastened a line the
same length as the poles to the small ends, and on the other end of
the line I tied a small hook and fastened on a shot sinker. The bobbers were clipped on anywhere to adjust the depth, a feature that
we found very important, for certain fish will often bite at one depth
when they ignore bait that is deeper or shallower. We found these
poles very handy in a number of ways. We could start fishing any-
time with a minimum of rigging, and when we finished fishing at
one place we only needed to discard the bait that was on the hook,
catch the hook in the butt of the pole, tie the poles on our car-top
carrier, and we were ready to move on to the next place. They required no special skill or practice to use—and the rankest amateur
could handle these poles as gracefully as a 'professional—and they
turned out to be real fish getters.

Our trip lasted for thirty days, but we didn't spend that whole
month fishing. Travel took considerable time on so long a trip, and
we spent a great deal of time digging many species of clams, crabbing, catching scallops, gathering oysters, and collecting sea urchins,
and more time exploring, observing, swimming, or just plain loafing.
We gathered edible seaweeds and seaside plants, picked wild berries
and other edible wild plants, cooked strange dishes over campfires
and carried on long conversations with local people we met on the
way. Our route hugged the coast through ten states and one Cana-
dian province, and we passed through five of those states on our
way back to our home in Pennsylvania. We studiously avoided all
places where we had been before, so we approached each stretch of
shore with no knowledge of its fishing or foraging possibilities. We
probably spent no more than thirty hours of the entire month at
actual fishing, and yet in that month we caught more than six hundred fishes of many shapes, sizes, and species. We could have eaten

fish at every meal, if it hadn't been for the presence of so many other kinds of delicious seafood that could be had for the taking. We often gave our catch, dressed, to some fellow camper whose fishing luck had been bad. We also salted, pickled, and smoked some fish to take home with us. Anyone can do as well who will use common fishing sense and simple equipment, and not waste his time going after the spectacular kinds of fish that are better for boasting than for eating.

THE TINY SMELT
Omerus mordax

Since we are talking about the small, sweet-fleshed fish that swim close to the shore, let's talk about the smallest and sweetest of them all, the Smelt. While traveling in northern Maine I met a fellow camper who was one of the best seaside foragers I have ever known. One day when my wife didn't want to go fishing, I went to his camp and he suggested that we catch a mess of Smelt for dinner. How a man who wasn't a native could be sure of naming his fish and then going out and getting it I didn't know, but I agreed to go along if only to see him discomfited. We drove to a sardine cannery and after securing permission, which was graciously given, we fished from the cannery wharf. We used very small hooks, baited with the tough foot muscles of Dog Whelks. No sooner had we dropped our lines into the water than they were taken by 10-inch Smelts, and we began pulling them in one after another. I asked my friend how he knew that this particular fish would be found in just this place and he, without taking his eyes from his dancing bobber, answered, "I smelt 'em."

I protested that his pun "smelt" much worse than the fish we were catching, but his facetious answer had much truth in it. The wharves of fish canneries do not smell exactly like attar of roses, but the fish scraps that fall into the water below furnish continuous chum that attracts many fish, including Smelts, making these ideal spots for wharf fishing. When seeking Smelt in New England, just follow your nose.

The Smelt is a slender, silvery fish, and adults average only about 8 inches in length. If you catch a 14-incher you are approaching the record. The drawing will give a good idea of general shape and fin arrangement. The Smelt can be distinguished from other small, silvery fishes by its mouthful of well-developed teeth. These teeth enable it to feed on smaller fishes and crustaceans. On the East Coast they range from the Gulf of Saint Lawrence to Virginia, but Smelts love cold water, and in the southern part of their range they are found near the shore only in winter. They are also found along the West Coast and in the Great Lakes. There are landlocked forms in many northern lakes, for the Smelt is not exclusively a saltwater fish, although it seems to me that those from the cold saltwater of the Maine coast are better flavored than the freshwater forms.

In late winter or early spring they enter streams to spawn, and this is when the famous "Smelt runs" occur. Once, in the state of Washington, I saw a Smelt run where there actually appeared to be more Smelt than water in the stream. State law limited each fisherman to 20 pounds, and that day everybody on the stream caught the limit. Several golfers from a nearby course, using their golf bags for nets, dipped up 20 pounds of Smelt apiece. Another man was using an inverted bird cage, and a sailor caught the limit using only a bucket and his little white hat. Nearby residents were laying in a supply of Smelts with colanders, kitchen strainers, flour sifters, or just ordinary pots and pans.

Such dense runs of Smelt are rare nowadays, but they still occur sporadically. When I hear old-timers talk about the plentiful Smelt of their youth I suspect they are talking about that one good year that occurred away back when—and that such runs did not occur every year, even then. I know that Smelt have been fished to destruction in some areas, but when these little fish have access to the sea, they gain a period of respite that allows them to recover, and then one day everyone is surprised by a dense Smelt run in streams where they have been spotty for years.

Despite its small size, the Smelt refuses to take second place to any fish. A few years ago, the Commonwealth of Pennsylvania stocked some of its freshwater lakes with Smelt as an additional food supply for the bass, pickerel, and pike they were almost hand-raising for the sport fishermen of that state. When the Smelt became plentiful, someone discovered that they would bite through ice. Now, in winter, these lakes are dotted with ice-fishermen.

One secret of the Smelt's general popularity is its willingness to bite enthusiastically on a variety of baits. The Smelt will never furnish the big thrills of fishing, but it can deliver a great number of small thrills in a very short period of time.

When wharf fishing for Smelt, use the bamboo-pole rig with a No. 8 or No. 10 hook, and bait with the feet of small Dog Whelks. They will also bite on worms, clams, or cut herring, but the feet of Dog Whelks are tough enough to stay on the hook, fish after fish, and this enables you to spend your time fishing rather than baiting. (You can't catch a Smelt, or any other fish, while your hook is in your hands instead of in the water.) I once caught 23 Smelts on one Dog Whelk bait. As you remove each Smelt from the hook, pinch its head hard, just behind the eyes. This will crush the brain, killing the fish instantly and mercifully.

The bamboo-pole rigs proved their superiority that first day of Smelt fishing. Another fisherman on the same wharf was using an expensive spinning rig and clam bait. He, too, was catching Smelts as fast as he could bait up, cast out, hook the fish, reel it in, remove it, rebait and cast out again, but all this ritual took so much time that my friend and I were each catching three fishes to his one. He soon saw the advantage of our tough baits and borrowed a few Dog Whelks from us, but handling that fancy rig took so much time that my friend and I continued to catch two fishes to his one.

The clip-on bobber is very useful when fishing for Smelt, since it can be adjusted for depth. Smelt bite best at four to eight feet below the surface. The sensitive bobber also lets you see every bit of the action from the time a fish first starts fooling around the bait. The moment the bobber dips under, a slight jerk will hook the fish, then it can be brought to the surface and landed. Too hard a pull will tear the hook from its soft mouth.

When I stated earlier that my wife and I caught more than six hundred fishes with not more than thirty hours' fishing, I should have said that we could have caught a great many more if we had had any use for them. I could easily have caught six hundred Smelts from the fishery wharf in one day, if I had wanted so many. I caught more than a hundred in one hour, and that's pulling them in at a rate of almost two per minute.

On arriving back at camp, I clipped the heads and tails from the little Smelts with a scissors. After this was done, I ran my fingers down the sides of the fish, from the tail toward the head, to press out the viscera. Even a novice can clean ten to twelve fish a minute this way. There are those who say that a Smelt should be cooked entire, with no cleaning whatever, but I prefer them with the head and entrails removed. Not only do they taste better, but they will then lie flat and straight while frying instead of curling up like corkscrews.

We put a huge iron griddle on the campfire, greased it, lightly salted the cleaned fish, and cooked them with no coating whatever. They were superb. They are also very good dipped in egg and rolled in cracker crumbs, or dipped in an egg batter, before frying.

Since we couldn't possibly eat all the fish we had caught, we made them into Pickled Smelt, a noble dish. The cleaned fishes were placed in a strainer and set over boiling water to steam just 3 minutes, barely enough to firm the flesh by curdling the protein, then they were stacked neatly into a deep glass dish and covered with a pickle made by boiling together for 5 minutes the following: 4 cups vinegar, 6 Bayberry leaves, 2 level tablespoons salt, and 1 tablespoon mixed pickling spices. This mixture was then poured boiling hot over the fish. Three days later we ate the Pickled Smelt on an assortment of fancy crackers, accompanied by beer for the adults and lemonade for the children. They were fine.

THE PLENTIFUL POLLACK
Pollachius virens

The Pollack is the most active fish of the cod family and the one most available to wharf fishermen. From Cape Cod to Labrador, there is hardly a wharf, jetty, or rocky point that will not yield enough Pollack for a good meal. Other fish may be localized—one place being good for Smelt fishing, another yielding flounder, while still others may be the best places for mackerel—but everywhere you will also find the Pollack, usually outnumbering the more desired fishes ten to one. Not that the Pollack isn't a good fish—it is a fine food fish when properly prepared—but any fish as plentiful and easily caught as the Pollack soon comes to be considered unattractive by bored local fishermen.

A group of us once were fishing from a cannery wharf on Mount Desert Island, in Maine, where hundreds of these fish were biting freely. I soon became tired of such easy sport and made a bet with a bystander that I could catch one with a bare hook, using no bait whatever. I kept the hook moving about some four feet under the water and within ten minutes I had caught, not one, but six good-sized Pollack by this crazy method.

There are records of Pollack three and a half feet long and weighing thirty-five pounds, but I've never seen one that weighed as much as ten pounds. Most of those caught from wharves weigh from one to four pounds. They resemble other cods in having three dorsal and two anal fins, but the barbel on the chin which characterizes most cod is so minute on the Pollack as to be unnoticeable. They are scaleless and blue-black in color. I usually use Dog Winkle as bait because it is tough enough to stay on the hook for many fish, but the Pollack will bite on anything or nothing when running thickly, as I've observed.

Use a clip-on bobber to hold the hook four to eight feet below surface and keep your eye on the bobber. If the Pollack are there, the bobber usually starts dancing immediately. The moment it ducks under or starts skidding across the surface, set your hook with a slight jerk, play the fish to the surface and lift it onto the wharf. That's all there is to Pollack fishing.

Pollack is full of small bones, so should be filleted. The bones are localized, and it is possible to get two reasonably good-sized fillets from each Pollack. Slip a sharp-pointed knife under the skin and make a slit down each side of the fish, just back of the gills. Now make two slits through the skin, down the fish's back, from just behind the head to the tail, one slit on each side of the dorsal fins. Peel the skin back until the flesh on the upper part of the fish is exposed. Cut the flesh loose from the rayed backbone by letting your knife slide just outside these bones along the back, but do not cut deeper than the median line along the fish's side. Now cut along this median line and remove the fillet. Turn the fish over and repeat on the other side. Don't expect too much, just two slender fillets from each fish, but it is really over half of the edible meat on each Pollack, and is all the flesh that can be removed perfectly boneless from this fish. If you want more fish catch more Pollacks; it is easily enough done. When a Pollack is cleaned in this manner, one never need go through the process of beheading, skinning, eviscerating and other messy operations that make most fish cleaning an irksome task. One merely discards the head, skin, bones, entrails, and other offal in one piece with the unused portion of fish. The

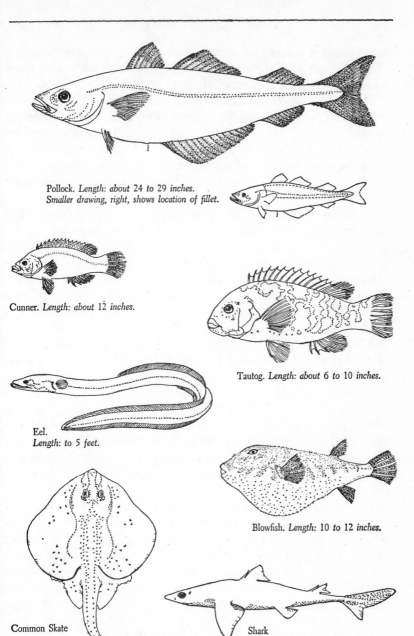

Pollock. *Length: about 24 to 29 inches.*
Smaller drawing, right, shows location of fillet.

Cunner. *Length: about 12 inches.*

Tautog. *Length: about 6 to 10 inches.*

Eel.
Length: to 5 feet.

Blowfish. *Length: 10 to 12 inches.*

Common Skate

Shark

accompanying drawing will give you the shape and fin arrangement of the Pollack, and the dotted lines on the side show the location and relative size of the boneless fillets.

Some people complain that Pollack doesn't taste fishy, and some rejoice in this fact. The fillets are excellent fried, boiled, baked or broiled. To bake, wrap the fillets in aluminum foil and cook either over an outdoor fire or in the oven. You can wrap several fillets in the same piece of foil and include a savory stuffing, or wrap each separately in a slice of bacon before putting it in the foil, or just bake it plain.

Pollack is not strongly flavored, so be sure to season it before cooking. Grind one large onion in the food chopper and save all the juice. Mix the onion mush with two tablespoons salt, two tablespoons vinegar and one teaspoon mace. Let the fillets marinate in the mixture for one hour, then rinse in cold water and fry, bake or broil.

To boil Pollack, tie half a dozen large fillets in a piece of cheesecloth, lay on bottom of a kettle and cover with two quarts water, one tablespoon salt and the juice of one lemon. Bring quickly to a boil then simmer for five minutes. Serve hot with dill sauce or anise sauce.

Corned Fish is one of the simplest ways to preserve a surplus for a day or two. It can be made with any fish but is especially good when made of Pollack. Cover each fillet with as much salt as will cling to it, stack the fillets in a deep dish, and leave in the refrigerator for four hours. The salt will extract liquids from the fish flesh and you will find the fish covered with brine. Drain this off and rinse under cold running water, then wipe dry. Keep in the refrigerator until used, preferably within three days. Commercial fishermen, who like their fish plain and unadorned, like corned fish as a breakfast dish. As many corned fillets as are needed for a meal are tied in a piece of cheesecloth, placed in a kettle and covered with cold water. This is put on high heat and as soon as it reaches a simmer the water is poured off and the whole process repeated. The second time it reaches a simmer the heat is turned down a bit and the fish allowed to cook for five minutes more. The fillets are then served whole, each piece seasoned with a dash of freshly ground black pepper and a pat of butter. This fish is still on the salty side so it should be served with unsalted potatoes.

In Hawaii, I learned a way to dry fish fillets with the aid of the sun. Direct sunlight alters the taste of fish to impart a new and distinctive flavor. Prepare the fillets exactly as one does for Corned Fish, but as soon as the salted fillets are washed and wiped dry they are placed in the direct sun, inside a screen cage to keep the flies from them. They should be brought inside at night and placed outside again next day. About two days of sun imparts the best flavor, and then they should be cooked immediately or kept in the refrigerator until used, preferably within a week. These sun-flavored fish are excellent broiled or fried. In Hawaii, we ate them with poi, but I have found them very good with unsalted potatoes or hominy grits. When I first tasted Hawaiian sun-flavored fish the flavor was so surprisingly different from that of ordinary fresh fish that I wasn't sure whether I liked it or not. However, as was the case with many Hawaiian foods, the more I ate the more I wanted.

CUNNER AND TAUTOG
Tautogolabrus adspersus and Tautoga onitis

These two fishes belong to the Wrasse family, which are chiefly tropical fishes, and yet strangely enough these two love cold water and reach their greatest development in the northern part of their range. The Cunner occurs from Labrador to New Jersey and can be caught by dozens from wharves, jetties, and rocky shores throughout New England. It is the bitingest fish of them all, savagely attacking almost any bait with a series of incredibly fast jerks that are transmitted up the line almost like an electric shock. The secret of catching them is to use a very small hook and a small bait floated so it just clears the bottom. A large bait will be nibbled without the Cunner ever taking it into its mouth. I like to stand on a rock at low tide and fish for them in five to ten feet of water. The schools of Cunner are visible through the clear water, and by watching the tiny white Dog Whelk's foot that I use for bait, I can see the exact instant that it disappears into the maw of a fish and set the hook with split-second timing. Then it is a mere matter of keeping a tight line on the fish, working it to the surface, and then lifting it ashore. If fishing is slow I toss in a few soft clams that have been chopped finely, their shells as well, and the Cunners soon swarm around ready to bite anything I offer—clams, worms, shrimp, squid, dog whelk, bread balls, or any other kind of bait. When I go clam digging I always save the clams on which I accidentally break the shell to use as this kind of chum. They attract not only Cunner but other fish as well.

The Cunner is not a very prepossessing fish. It has been known to reach a length of fifteen inches and a weight of two and a half pounds, but I've never caught one more than a foot long nor heavier than one pound. They are easily recognized by the single long dorsal fin shown in the illustration. The color is a rusty brown shading off to brownish-yellow in the belly region. The skin is tough, and the best way to prepare the Cunner for cooking is to skin it completely and cut off the boneless fillets. Make an incision around the rib case then remove the fillet by cutting down from the back. Turn the fish over and take the fillet from the other side. These fillets can be cooked by any fish recipe and when very fresh they are excellent.

An Englishman I took Cunner-fishing one day complained that he couldn't get fish and chips in this country. He was dining with us, so that evening I made "chips" by slicing good Maine potatoes end-wise into rather larger pieces than we usually use for French-fries. I cooked them to a light, golden brown in deep fat. Then I made a batter of 1 cup flour, ½ teaspoon baking powder, 1 teaspoon salt, ¼ teaspoon black pepper, 2 beaten eggs and ½ cup water. Each Cunner fillet was dipped in the batter, dropped into the hot fat and fried to a turn, then drained on a paper towel. After the first bite of this battered fish my friend apologized for having once said that the Cunner seemed a dull fish.

The Tautog, or Black, as it is called in some sections, is a sort of big brother to the Cunner. It has the same single dorsal fin reaching from back of its head almost to the tail. It can be distinguished from the Cunner by its larger size and much darker color. There is a record of a Tautog thirty-six and a half inches long, weighing twenty-two and a half pounds, but an eight-pounder would be something to write Uncle Gus about, and most of the Tautogs you catch will range from one to three pounds. They can be caught very handily on the bamboo pole rig, but you must be on your toes. The Tautog lies under cover, darts out and takes a bite of your bait then scampers for cover again. As one fisherman told me, the right time to set the hook is the split second *before* you feel the Tautog bite. If you have a well-developed precognition you should have no trouble catching all the Tautogs you can use. Fish for them around rocks, wharves, old wrecks, piling, or weed patches, wherever there is adequate cover, for the Tautag doesn't like to venture into the open.

Like the Cunner, the Tautog should be skinned and filleted, but in this case the fillets are considerably larger. Cook it in any way that you use fish fillets and you will not be disappointed in the flavor. One recipe that I have found delicious when made with these fillets is a variation of a Hawaiian fish dish called the *laulau*. I bake the fillets in aluminum foil instead of the *ti* leaves Hawaiians use and invoke nothing that is not locally obtainable. Buy a fresh coconut from the market, remove the meat and grate finely. If you can't get fresh coconut, canned or grated coconut that is packed moist will serve as well. Stir 1 cup boiling water into the grated coconut meat, and press and knead with the fingers for about 10 minutes. Then put the coconut in a jelly bag or a double thickness of cheesecloth and squeeze the liquid from it. You will get about 1½ cups of a thick, creamy, coconut-flavored liquid. Cut squares of aluminum foil and fit the center of each piece into a shallow bowl. In each piece of foil put enough Tautog fillets to serve one person, dusting each fillet with salt and pepper, pour on ¼ cup coconut cream and add a few celery leaves from the top of a bunch. Gather up the corners of the aluminum foil and twist to make a tight package. When there are as many packages as there are people to be served place them in a bake pan without water and bake at 350° for 30 minutes. Serve the packages directly on plates and allow each diner to unwrap his own, for the heavenly aroma that arises from a *laulau* as it is opened is definitely part of the treat.

EELS
Anguilla rostrata

I could never qualify as an expert Eel-fisherman for I never caught an Eel on purpose in my life, although I have caught hundreds while angling for other kinds of fish. The only time I ever deliberately tried to catch an Eel was during a summer vacation on the banks of a tidal stream in New Jersey. I fished this stream for a fortnight for white perch, striped bass, drum and weakies, but all I seemed to catch were Eels. I consider Eel a great delicacy and prepare it in many ways, but that year I had already tired of Eels and toward the end was throwing most of them back. Near the end of my vacation, some friends visited us from the city and wanted proof of my avowed prowess as an Eel fisherman and chef. I assured them that catching enough Eels for a gourmet dinner was no trick at all. We set out, drifting on the tide, to make our catch. We caught dozens of nice perch, several good-sized drum, a few small sea bass, many sculpins or oyster-crackers, and one member of the party landed a small, but beautiful, striped bass, but we never caught sight of an Eel. We used the same rigs and bait on which I had been catching Eels every time I went fishing, and yet, on this one day when I really wanted them to bite, not a single Eel took bait. I took a good ribbing, and invited everyone to go out next day to try for perch and bass. We caught a few perch, more crabs and one or two small stripers, but we also caught a huge string of Eels. Since then I will never admit to anyone, not even myself, that I am going Eel-fishing. I fish for other species in waters where Eels abound and I never fail to catch all the Eels I can use.

I have caught Eels from Maine to Mexico, and their total range in America extends from the St. Lawrence to Brazil. No one will have difficulty in recognizing this long, snake-like fish. Eels can reach a length of 5 feet, and where I usually fish, 3-foot specimens are not uncommon. The Eel really has only one fin, a continuous dorsal-caudal-anal fin that fringes the after half of its body. When caught in fresh or brackish water the skin is heavily pigmented, nearly black. The place to catch Eels is in brackish tidal streams everywhere. In such places Eels can also be caught in eel-pots or eel-fykes, which are traps made of wire, lath or netting with narrow-mouthed, funnel-shaped openings, but since I always catch as many Eels as I want as a by-product of other kinds of fishing, I have never bothered with these devices.

The Eel has one of the strangest life-histories of any creature that swims. After reaching maturity in rivers, streams, tidal inlets and bays, the Eels is suddenly seized with an urge to return to the ocean. From the *burns* of Scotland, the *kills* of Holland, and from the streams of all Europe and America, they all come to congregate for spawning in one area of the South Atlantic between the Bermudas and the Bahamas. Here they lay tiny eggs by the millions, then all the adult Eels die. The minute larvae that hatch from the billions of eggs are very un-eel-like in appearance and once were thought to be a separate species. They are called *leptocephalus*, and they are tiny things, cigar-shaped in outline but thin as paper. For two years they travel toward the streams from which their forebears came, some heading for Ireland, some for Scandinavia, some for Texas, and others for all the other stream mouths and brackish inlets that ring the North Atlantic. By the time their unerring instinct leads them to the identical stream from which their mothers came, the little Eels have graduated from the *leptocephalus* stage and are shaped much like adults but they are only 2 to 3 inches long and as thin as coarse knitting needles. They are almost transparent until they enter brackish water, but there they soon darken until their skins are almost black.

During their immense journey from their tropical birthplace to their ancestral streams, these little Eels are preyed upon by everything that swims, flies, or crawls along the bottom. By the time they reach our shores they have been reduced in numbers by a ratio of many thousands to one. And yet, it is very hard to imagine more Eels than the number that sometimes enter a small stream in the spring. I once stood at the juncture of a small stream and salty bay in Connecticut, and saw baby Eels crowd in that stream mouth from the bay by millions. A bucket dipped at random would have contained a greater weight of Eels than water. And still the decimation of their numbers went on. Sea gulls were gobbling them down, a dozen at each gluttonous bite. Mackerel and stripers had followed them from the sea into the bay and were harrying the outer edges of migration. On the bottom, crabs of several species were growing fat on this abundant windfall of food. I, too, dipped in a strainer, then shook the Eels I had caught in a paper bag with some seasoned flour, and fried them over my campfire. They were surprisingly delicious.

The adult Eel has firm white flesh with a delicious flavor, more reminiscent of fowl than of fish, and yet this fine food fish is very unevenly appreciated throughout its range. It sometimes seems that wherever Eels are scarce and hard to catch they are eagerly sought and considered a delicacy, but where they are abundant and easily caught they are disliked. One exception to this rule seems to be among the real old-timers around Narragansett Bay, the descendants of the early settlers of Rhode Island. The pioneer farmer-fishermen, who settled this beautiful land, early learned to appreciate the huge Eels that could be caught in these waters and developed many delectable Eel dishes. Some of their descendants inherited the tradition and still set eel-pots or go Eel-fishing. There was a derogative saying in colonial America that all the Rhode Islanders asked of life was a jug of rum and a string of Eels. They could have wanted worse things.

Most people like Eel from the first taste, if it has been skillfully prepared, so the only possible objection one can have to this fine food fish is its serpent-like appearance, and this is not discernable by the time it reaches the dining table. Perhaps the remnants of prejudice against this delicacy will be destroyed by the fact that several exclusive French restaurants along the Eastern seaboard have recently begun featuring delicious Eel dishes at fantastic prices.

As I have indicated, Eels are easy to catch, and I think most tidal creek fishermen will agree that it is much easier to get an Eel on your hook than it is to get it off. When you have struggled for half an hour trying to hold a slippery, squirming Eel with one hand while trying to disgorge the hook with the other you will understand why so many fishermen dislike catching Eels. To avoid this messy business, carry a hook-disgorger when fishing in Eel waters. This is no more than a stiff wire bent into a hook at one end, with a handle fastened to the other end. In use, slip the hook over your line or leader, slide it into the Eel's mouth until the hook is engaged. Now hold the line with one hand and the handle of the disgorger with the other, letting the Eel hang free. Move your hands in a circular motion whirling the Eel over and over your line. By the time the Eel has made the third circuit it will usually come free. Turn your back to the water while you are doing this or there is a good chance that you will throw your Eel back in the creek when it comes free. In a boat, it may land in someone's lap.

To skin an Eel, cut through the skin all around the Eel just back of the head. Grasp the skin with a pair of pliers, hold the head in one hand, and pull off the skin with the other. It needs a hard pull, but once the knack is acquired you can skin rapidly. Slit the belly from vent to throat, remove the intestines, cut off the head, and your Eel is cleaned.

Many fine chefs consider Eel an essential ingredient of Bouillabaisse (page 87 ff.), but I dislike finding 2-inch chunks of Eel with the bone still in it in my Bouillabaisse, so I first fillet the Eels when using them for this purpose. Cut the cleaned Eels into 6-inch lengths and chill them for an hour or so in the refrigerator. This firms the meat and makes it much easier to remove from the bone. With a small, very sharp knife make a cut from the Eel's back, right next to the backbone, down to the rib case. Then, holding this cut open with

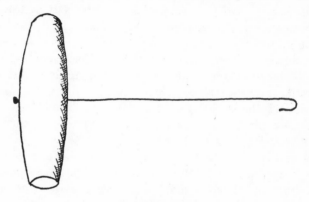

Hook-Disgorger

your thumb, cut the fillet loose from the rib case, using a series of shallow cuts made with the very point of the knife. Cut the fillets into 1-inch pieces and they are ready to be added to your Bouillabaisse with the other filleted fish.

Browned Eel with Onions is a fairly simple method of cooking Eel that delights everyone. Slice 12 medium-sized onions and 1 green pepper and fry them together in ⅓ cup cooking oil. When onions are translucent, drain and keep hot. Cut the cleaned Eels, about 2 pounds, into 2-inch sections. Dip in milk, sprinkle with salt and pepper, then roll in fine, dry bread crumbs. Fry in the same fat in which the onion and pepper were cooked until golden brown. Unlike most fish, Eel should be cooked slowly and thoroughly, so fry it just as you would fry chicken. Put the onions and pepper on a platter and arrange the pieces of browned Eel on top. Serve with a good helping of hominy grits, or better yet, a generous serving of wild rice. A glass of thoroughly chilled Chablis or Moselle will be just right with this dish.

Eel Salad with Herb Dressing is an elegant dish for which you would pay outrageous prices in a good restaurant. The tidal creek fisherman who has learned to use the Eels which he unwillingly catches, instead of cursing at them and throwing them back, can make this epicurean dish in his own home and it will be as delicious as that made by the best French chef who ever boiled an Eel. You need about 2 pounds of cleaned Eel cut into 3-inch lengths. Finely slice 1 onion, 1 carrot and 1 clove of garlic in the bottom of a saucepan. Put the Eel on top of this vegetable bed and add 1½ cups dry white wine, 1 teaspoon salt and about 10 drops Tabasco sauce. Cover; bring quickly to a boil then lower the heat and simmer for 30 minutes. Let cool, then remove the Eel pieces and refrigerate them until they are thoroughly chilled and quite firm. The onion, carrot, garlic and wine were merely to flavor the Eel and can now be discarded. Pick the chilled Eel meat from the bones. It will taste something like firm crab meat and can be served in a number of ways. For an unusual seafood cocktail, line the glasses with the fine heart leaves of lettuce, fill with diced Eel meat and cover with Herb Dressing. For a hearty luncheon dish that is just right for a hot summer day, make a bed of shredded lettuce, heap the boneless Eel over it and cover with Herb Dressing. Or, you can use the boneless Eel in any salad recipe that calls for crab, lobster, shrimp or cold chicken. It will not taste exactly like these meats but it is guaranteed to be delicious.

To make the Herb Dressing that goes so well with cold Eel, combine ¾ cup salad oil, ¼ cup white wine vinegar, 1½ teaspoons salt, ¼ teaspoon freshly ground black pepper, ¼ teaspoon dry mustard, 3 tablespoons each of chopped fresh chives, parsley and dill, 1 teaspoon dry basil and 1 teaspoon dry marjoram. Make this dressing about one hour before it is to be used and allow to chill in the refrigerator. Shake well before using.

THE BLOWFISH OR NORTHERN SWELLFISH
Spheroides maculatus

The Blowfish belongs to an essentially tropical family but this one species breeds as far north as New York and is sometimes caught on the Maine coast. It likes sandy shores and is often the easiest fish to catch along the New Jersey and eastern Long Island shores during the summer months. An average Blowfish is only ten to twelve inches long, but this is a blocky, squarely-built fish with considerable bulk to its length. It is grayish-green or brownish in color with lighter undersides with about a dozen irregular darkish spots along the edges of the light underside. Its scales have been modified into rather small prickles. When threatened or handled this fish has the peculiar habit of inflating itself with water or air until it is blown up like a basketball and almost spherical in outline. When blown up in this manner the prickles stand out and probably give the fish some protection. When a Blowfish is caught it can be made to inflate itself by stroking the loose skin in the belly region.

When Blowfish can be caught at all, they can usually be caught in large numbers, for they tend to congregate in schools. An unusual sight is to see a large number of Blowfish "ganging up" on a Blue Crab. A single Blowfish will never attack a full-grown Blue Crab but when there are a number of them together, some of them will engage the crab from the front while others run in from behind and bite off its legs, soon reducing it to helplessness, and then all join in the feast and finish the crab in short order. This is an amazing performance, for it implies a degree of cooperation and reasoning power which I find very hard to assign to such a stupid-looking fish. The teeth are fused in a beaklike structure that is powerful enough to crush a crab shell or the finger of a careless fisherman.

Blowfish will bite on worms, squid, clam, cut fish or many other baits but their favorite is shedder crab. On one long seaside journey, my wife and I stopped by Island Beach State Park in New Jersey. The elaborately outfitted surf fishermen smiled kindly at our crude fishing equipment and told us there was no need even to put a hook in the water as nothing was biting but Blowfish. They couldn't know that Blowfish was exactly what we wanted. I rigged our handlines with three small hooks apiece and baited them with joints of shedder crab. We had no sooner tossed them out than we began pulling in nice Blowfish, one after another. In an hour we had caught thirty-six. The only secret to catching them was the use of very small hooks and small baits, for the Blowfish has a small but powerful mouth. Handlines such as we were using will catch as many Blowfish as the most expensive equipment made.

On another occasion, we were walking along a little sandy bay in the month of May. I happened to look out in the shallow water near the shore and saw several dozen large Blowfish swimming around. In a short while I had scared up some bait and rigged our bamboo poles. We waded out to where the water was knee-deep and the Blowfish came swimming all around us. Catching them was a mere matter of dropping a baited hook in front of one after another. In a short while we had caught as many as we could use.

One needs to catch a considerable number of Blowfish to make it worthwhile, for the proportion of edible meat is rather small. All the meat is in one piece, lying on either side of the central backbone. The only bone in the Blowfish is the backbone; even the ribs have degenerated into mere bumps along either side of the spine. To clean a Blowfish, simply cut through the skin and backbone just behind the head, slip the end of the meat-bearing spine out through the hole, grasp it tightly and pull it out by turning the skin inside out. It will skin all the way back to the tail and even skin out the tail itself. You will be left with a drumstick-shaped piece of meat with the skinned-out tiny tail on the small end of it.

Although formerly considered a "trash fish," to be thrown back when caught, the Blowfish is more and more being recognized for the superior delicacy that it is, and now the little drumstick-shaped pieces of firm delicious flesh command premium prices in fancy fish markets under the euphemism of "sea squab." I have mentioned using Blowfish in Bouillabaisse (page 91). For this purpose, the meat is sliced from the backbone and used exactly as other fish fillets. As fried or baked fish the bone is not bothersome and the fish is eaten in the same manner as one eats a chicken leg. To fry Blowfish, dip the pieces in an egg beaten with a little cold water, roll in fine bread crumbs and fry to a light brown. To bake it, cover the top of each piece with a short strip of bacon, set it in a 350° oven and bake until the bacon is crisply done. As soon as you have tasted it I am sure you will agree that Blowfish is a fish dish worthy of being served to the most finicky of epicures.

SHARKS

By far the most plentiful shark on our coasts is the Dogfish, *Squalus acanthias*, also known in various localities as the Greyfish, Spiny Dogfish, Cod Shark, Sand Shark, Thornback Shark, Bone-dog and Skittledog. This is the gray, rough-skinned fish that is familiar to every college student who ever studied comparative anatomy, for they must dissect foul-smelling specimens preserved in formaldehyde and memorize every organ, every muscle, every vein and artery, and each part of its cartilaginous skeleton. The sandpaper-like skin is slate gray on the back and light gray or whitish on the belly. It is neither a large nor a dangerous shark, the very largest being about 4 feet long and weighing twenty pounds while the average that will be caught will weigh between six and ten pounds. It has a rounded head and a flattened snout, and its mouth is well-equipped with small, sharp teeth bent toward the outer corners of its mouth. There are usually twenty-eight teeth in the upper jaw and twenty-four in the lower. It has two dorsal fins and just ahead of each one of them there is a sharp-pointed spine. If the freshly caught dogfish is carelessly handled with the bare hands it may inflict a painful wound with these spines while threshing about. This is the only danger from the Dogfish, for although stories of its attacking swimmers are common, none of these tales has ever been authenticated.

On the Atlantic coast, the Dogfish is found from Nova Scotia to Cape Hatteras, and, on the Pacific, it is found from Alaska to Northern California. These sharks hunt in packs and eat crustaceans, mollusks, other fish, jellyfish, and nearly any animal matter they can find be it dead or alive. The Dogfish gives birth to living young and has from one to ten young in each litter. These "pups" are from six to ten inches long, depending on the size of the mother. They are able to shift for themselves as soon as they are born, and newlyborn Dogfish have been observed attacking herring as large as themselves.

During World War II, when the imports of cod-liver oil from Norway were suddenly cut off, drug companies began searching for an acceptable substitute and soon discovered that the livers of Dogfish were even richer in vitamin A than those of cod. There followed what I have always thought of as the great Dogfish boom. Competitive bidding by the drug companies soon ran the price of these oil-rich livers to several dollars per pound. As any Dogfish of more than six pounds will have more than a pound of liver, the Dogfish suddenly became a very valuable piece of merchandise.

I was living in Seattle at the time and the harbor was crowded with Dogfish that hitherto had been considered nothing but a nuisance. While the boom lasted I spent every spare minute Dogfishing.

I used a heavy hand line with a wire leader and a stout hook, and after catching the first Dogfish on a piece of stewing beef I had no trouble getting bait, for they have no objection to eating one another, so cut Dogfish was as good a bait as any. Head and offal were thrown in the water where I was fishing and served to attract a constant supply of Dogfish to the area. Often the amount I received for the livers of the Dogfish caught on a weekend would be more than the salary I received for working the other five days of the week. While such conditions prevailed, it was very easy to persuade my wife that I should spend the weekends fishing rather than mowing the lawn or putting up the screens, and I lived in a husband's paradise.

It was during the shark-liver boom that I first tentatively experimented with the Dogfish as food. I had read that these little sharks were sold in European markets as "greyfish," so I knew they were at least somewhat edible. Baked, boiled, steamed or broiled they were barely edible, but when the flesh was cut in strips, dipped in batter and fried in the manner of English fish and chips, they were perfectly delicious, with a flavor reminiscent of lobster. A recipe for a suitable batter is given on page 232.

Later, in Hawaii, I found that shark meat makes excellent fish cakes. The easiest way to make this delicacy is to run the filleted flesh through a meat grinder, then while gradually adding a little salted water, knead and work the ground fish until it assumes a spongy texture. With dampened hands this spongy fish can be shaped into patties which are very good fried. For a change, add onion, green pepper or celery, one or all three, to the fish as you grind it. Others like to add a bit of cooked ham to the grind. The fish cakes should be fried in peanut oil and eaten with rice and soya sauce.

One can hardly fish for very long along either the Northeast Coast or the Northwest Coast of America without occasionally catching a Dogfish. Instead of condemning these little sharks as nuisances, treasure them as valuable food fish, no matter what local informants might say. Often I relieve fellow fishermen of the sharks they disdain. If you are timid about revealing the shameful fact that you are a shark eater, tell your companions that you are taking the fish home to bury around the sweet corn in your garden, as Squanto taught the Pilgrim Fathers to do. Later, you can invite these same fishermen to dinner when you are serving battered fried shark and bask smugly in the praise they will invariably heap on this same fish they disdained when it was alive.

RAYS AND SKATES

The Rays and Skates should be discussed together, as they are much alike and are used in exactly the same ways. As with the eel, I have never gone fishing for Skates or Rays, but I have caught a great number while angling for other fish, and they are always welcome aboard. It is not for their looks that I prize these monsters, for uglier fish would be hard to imagine. As you can see from the illustration, the Skates and Rays are flattened laterally with the sides extending into two winglike flaps with which the creature swims through the water much as a bird flies through the air. The one most commonly caught by the seaside or wharf fisherman is the Common Skate, *Rapa erinacea*, which is found on shallow, sandy bottoms from Halifax to Cape Hatteras. It seldom becomes more than 2 feet long, and lives on small crustaceans, mollusks, and other animal food. It is so distinctive in shape that no one will ever mistake a Skate for any other kind of fish.

After one experience, you will always know when you have hooked a Skate by the way it pulls. They are sluggish fighters, never jerking or running with the line. They simply sail about, resisting your upward pull by spreading their broad wings to enlist the resistance of the water. The experience is much like flying a kite upside down. When you have finally sailed your Skate to the surface and landed it on the shore or wharf, don't be repelled by its looks, for Skate wings are better than the finest fillet of sole. Kill the fish by a hard blow between its piggish little eyes, then cut off the "wings" as close to the body as you can. Just toss the rest of the fish overboard to attract other fish you would like to catch to the area.

The Skates and Rays, like other members of the shark family, do not have true bones, but have a cartilaginous skeleton that can be sliced with a sharp knife. The best way to clean a Skate wing for cooking is first to cut the wing into long strips about 1 inch wide, cutting through skin, cartilage, and all. Remove the skin from these strips, then slice the flesh from each side of the cartilage. This will leave you with two slender fillets from each strip. These can be cooked as you would use any filleted fish and they will be delicious, but the best way of all is first to cut the fillets into 1-inch squares,

then cook them exactly as you would scallops. You can use any scallop recipe from the scallop section of this book or any other recipe
you can find, and you will not be disappointed in the flavor of these
mock scallops. Some of my guests have insisted that they like them
better than the real thing. Try them the next time a Skate latches
onto your hook. All other species of Skates and Rays are edible and
all can be cooked in the same ways. Be sure you kill any Ray you
catch before handling it, for, like the Dogfish, they have sharp spines
that can inflict painful wounds on the careless fisherman.

One might think, from the foregoing parts of this chapter, that
the seaside forager or camper could expect to catch only the way-out
and unusual kinds of fish such as Skates, Rays, Eels and Blowfish,
but such is not the case at all. Most of the fish that will be landed
by the economically equipped casual fisherman will be perfectly
respectable, recognized food fish, but I expect the reader of this
book to know what to do with such a fish without any instructions
from me. If he doesn't, any stander-by (and there are often standers-
by when one is fishing) can instruct him, or he can get the information from any conventional cookbook.

I have caught mullet, hard-heads, croakers, bluefish, sea bass, stripers, kingfish, spots, flounder, fluke, and other well-known fish, besides the outlandish kinds emphasized in this chapter. In the upper
reaches of Chesapeake Bay, where the water is barely brackish instead of being truly salty, we even caught such freshwater forms as
black bass, crappie, and catfish. In brackish tidal creeks we caught
great strings of perch, both the white and the yellow kinds. All of
these were caught on the simple equipment described early in this
chapter. The sport fisherman may get his thrills in larger doses and
get more story-telling material, but with a minimum outlay for equipment, you can fill your frying pan with some of the best fish you
ever ate, and not have to devote much of your seaside time to fishing.
Good luck to you.

26. Edible Seaweeds

FEW Americans ever think of a seaweed as anything to eat, and yet you have probably eaten seaweed during the past week. Algin, a commercial product derived from seaweed, is widely used in the manufacture of ice cream, chocolate milk, and jellied puddings. It is this seaweed derivative that makes these desserts slide down your throat so easily and comfortably. It is also used in various pharmaceutical products, especially in hand lotions.

A few seaweeds appear in the dried state in our markets, but they find their chief sale among people newly arrived from other countries, and health food enthusiasts, and are almost unknown to the general public. In parts of Europe, seaweed is more highly appreciated than here, but it is in the Orient, in China and Japan, that this food really comes into its own. In Japan, there are seaweed farms in shallow coastal waters, and from all reports it is a very profitable type of farming. The farmer merely plants great bundles of bamboos on the bottom to furnish a surface where the seaweed can grow, and the sea does his planting, cultivating, and fertilizing; it is only necessary for him to return at harvest time and gather in the crop. The seaweed that is cultivated in this way is *Porphyra*, which also grows wild on our shores. We know it by the common name of Laver. The Chinese and Japanese residents of this country import considerable quantities of this seaweed from Japan, not realizing that it grows in abundance on both our Atlantic and our Pacific coasts.

While living in Hawaii, I learned to appreciate seaweeds as prepared by the Orientals. I enjoyed "long sushi," a kind of highly seasoned rice wrapped in seaweed. We often ate "limu," a local fleshy seaweed that gave a salty sea tang to our salads. Another great favorite was "kombu," a preparation of dried, pressed, and shredded kelp that has a sweetish taste.

Very little has been written about the edibility of our local sea flora. On delving into the lore in this area, I found plenty of references to the use of seaweeds in industrial products, and even in manufactured foods, but these require complicated processing. I did find a recipe for making blancmange from Irish Moss, and a notation that some people like to chew on Dulse, but beyond that practically nothing. I decided that if I wanted to learn to eat seaweeds, I must turn from the libraries and go to the seashore.

In one rocky cove on the coast of Maine, I found six kinds of seaweed I could use. On the higher rocks was an unlimited supply of *Fucus*, or Rockweed, which is used as a steaming agent in making both indoor and outdoor Clambakes. It is used not only for its mechanical properties, but also because it actually contributes flavor to the finished product. See page 107 for full directions on using this weed to steam some of the finest seafood you ever tasted.

In the tide pools were bright green sheets and ribbons of Sea Lettuce, *Ulva lactuca*, which is the largest of the bright-green seaweeds. It is nothing more than very thin, translucent sheets or ribbons of a bright green color. I have chopped Sea Lettuce and added it to a salad made of seaside wild plants, but I must confess I use it more for its pretty color than for its actual goodness. Fresh, it tastes good enough, but it must be chopped very fine, for it is tough and hard to bite. I know an epicurean vegetarian family who dry Sea Lettuce, then powder it and put it on the table in salt shakers. They use this dried seaweed as a salty condiment to flavor the marvelous tossed salads that make up a large part of their diet. I am no vegetarian, but I know of few houses where I would rather go for the salad course. Unfortunately, when Sea Lettuce is dried, it loses its bright green color, and the powdered product is black as black pepper. They tell me they dry it by merely laying it on clean wrapping paper in a warm attic room.

In a lower tide pool where the water was still surging in and out, I found the four other seaweeds that will be discussed here. The most

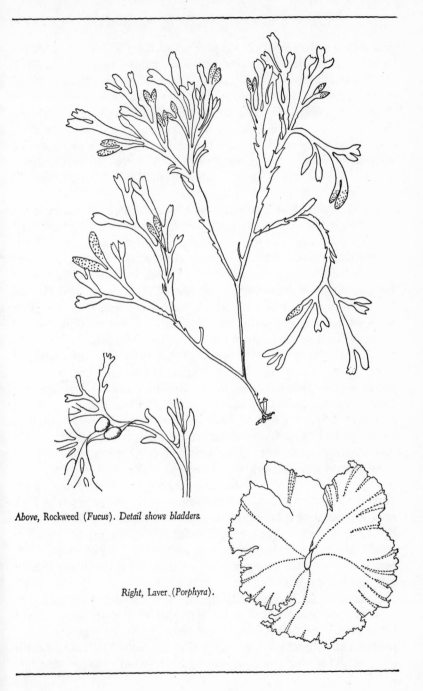

Above, Rockweed (*Fucus*). *Detail shows bladders.*

Right, Laver (*Porphyra*).

interesting of these is Dulse. I had heard that this is relished on the
coasts of Scotland and Ireland, and was anxious to try it. Known to
science as *Rhodymenia palmata*, it is found near low-water mark attached to rocks, shells, or even other seaweeds. There is a very short
stem that quickly broadens into a very thin, fan-shaped or tongue-
shaped expanse that is ribless and veinless and of a dark red or claret
color. The tongues or fans are sometimes further divided into flat
thumbs or fingers, and these again may be cleft into short, round-
tipped lobes on the ends. The whole plant is small, seldom over a
foot long.

The first time I tried to chew on a piece of Dulse, fresh from the
tide pool, I was greatly disappointed. It was tough and elastic, and
the sensation was very like chewing on a salted rubber band. However, when I hung some of it on a wire under my porch roof and
dried it for a week, it entirely changed, both in taste and texture. It
became pliable and tender, and to one who had learned to like the
limu and *kombu* of the Islands, it was savory and delicious. This
seaweed flavor is, like olives, an acquired taste, but it is not hard to
learn to like this salty, tangy sea flavor. The first time you chew on
a half-dried frond of Dulse you may find the flavor unpleasant, or
even slightly disgusting. If you persist, the second one will taste
better, and by the time you have chewed four bites of Dulse, taking
the bites a day apart, the task of learning to like this seaweed will
be replaced by the job of keeping a sufficient quantity on hand. The
dried product is not very attractive in appearance, being a dark,
greenish black with a whitish dusting of dried sea salt.

So popular was Dulse-chewing among the Irish, back when the Boston Irish were actually immigrants, that it is reported that dried
Dulse was hawked at railway stations and on street corners. The Irish
say that the flavor is only brought out by long chewing, which makes
this food unsuitable for eating at mealtime, but ideal for between-
meal nibbling while working. Subsequent generations of Boston Irish
apparently neglected to acquire the taste for dried Dulse, for I could
not find it for sale in Boston recently.

Dulse does not become crisp and brittle on drying as do the leaves
of land plants, but remains soft and pliable. I have found that I
could press a number of these soft pieces of Dulse together, like a
plug of chewing tobacco, and keep it in my refrigerator for a month
or more. This plug could be turned on edge and shaved into fine

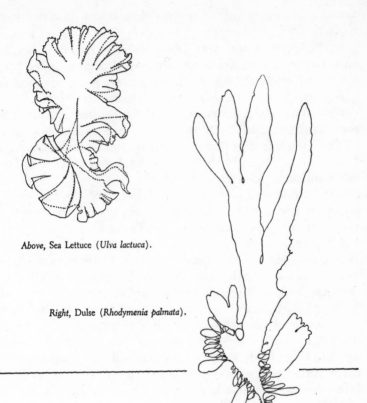

Above, Sea Lettuce (*Ulva lactuca*).

Right, Dulse (*Rhodymenia palmata*).

Right, Irish Moss
(*Chondrus crispus*).

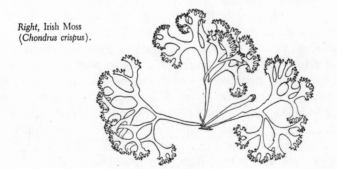

shreds with a sharp knife. Shredded dulse is very good in salads and slaws.

In the Mediterranean area, Dulse is used in cooked dishes. I once added some to a clam chowder and found that it cooked perfectly tender. I thought the flavor it imparted very good, but some of my guests, who had not acquired a taste for seaweeds, thought the chowder would have been better without it. On another occasion, I added some shredded, dried Dulse to a salmon loaf, and I *know* the flavor of that mundane dish was improved by this addition. I simply followed a recipe found in a regular cookbook, but added a half cup of finely shredded dried Dulse.

Another interesting seaweed of my tide pool was that same *Porphyra*, or Laver, that is raised by Japanese tideland farmers. It is found on both the Atlantic and the Pacific shores of our country, near the low-tide mark, on rocks, boulders, or even concrete piling. It is usually seen as thin frond that may be tongue-shaped, but is more often lobed, with a strongly ruffled margin, from a few inches to a foot long. It is red, purple, or purplish brown in color, with a filmy and elastic texture and a smooth, satiny sheen. After storms it is commonly cast on shore in considerable quantity.

Laver is more tender than Dulse and can be nibbled raw to give relish to other foods, but it, too, is better if half dried before being eaten. Either hung on a clothes line or spread on clean paper, it will be tender and chewable in a few days. This half-dried Laver can be carefully spread and stacked, one frond on another, and pressed into "plugs" in the same manner as Dulse. Pressed Laver can be kept in the refrigerator until used, preferably within four weeks.

Laver is the best of the seaweeds described here to use for cooking, and it makes a pretty fair soup. You can use freshly gathered Laver, or, even better, soak the dried product for 1 hour in enough water to cover it, then drain. Boil one cup of Laver in 2 cups of water, stirring frequently, until the seaweed is tender. Add one 10½-ounce can of beef consommé and the juice of half a lemon. Heat just to boiling, and serve with a twist of lemon peel floating in each bowl. This is best eaten with plain toast or unsalted pilot biscuits.

The thrifty Japanese make rice the center of each meal, and all meats, fish, savory vegetables, and other highly flavored foods are collectively known as *okazu*, a word that has no exact English equivalent but can only be translated as "a condiment to be eaten with rice." Laver makes a pretty good *okazu*. Put 3 tablespoons peanut oil in a saucepan, and brown 1 clove of garlic and 1 teaspoon fresh, green gingerroot. These are merely to flavor the oil, and are removed and discarded as soon as they are browned slightly. Put 2 cups of chopped, fresh Laver into the hot oil, cover the pan tightly, and sauté until tender, about 25 minutes. Add 2 tablespoons soy sauce and 1 teaspoon monosodium glutamate. Serve over rice. This can also be made of dried Laver, if you first soak it in fresh water for one hour.

The finest dish that I have developed so far in my experiments with Laver is Stuffed Laver Fronds. This was made of dried Laver, but I didn't soak the fronds first. I made the stuffing by combining 2 cups cooked rice, ¼ pound raw ground beef, ¼ pound mushrooms chopped, 1 medium onion, chopped, a little soy sauce, and a teaspoon monosodium glutamate. The dried Laver fronds were dropped into boiling water for about 1 minute, to make them soft and pliable, then spread out carefully. On each frond I placed about a tablespoonful of the stuffing, then rolled up the frond and placed it in a steamer. When all were stacked neatly in the steamer, I poured 2 cups boiling water over all, then set it on the fire and steamed it slowly for 50 minutes. These were really good.

Probably the most familiar of the edible seaweeds is Irish Moss, *Chrondus crispus*, which can literally carpet the sides and bottoms of the lower tide pools in rocky locations. One must see this plant in its native habitat to understand why it is called "moss." The individual plants look nothing like moss, but when it is thickly covering a rock it does look something like a bed of moss, except for the color, which ranges from olive-green through purple to almost black. The individual plants are from 3 to 6 inches high and about the same width, flattened, freely forking, subdividing several times until the terminal lobes are many and crowded. It is found at, or just below, low-tide level, usually attached to rocks, and is cast up on many shores, from North Carolina to Labrador. These cast-up pieces of Irish Moss are often white or cream-colored, because they are bleached by the sun.

Similarly, the Irish Moss purchased by housewives in grocery stores for making jellied fruit desserts and blancmange is usually a creamy white color. We are so used to this dead, pallid, sun-dried color that fresh, naturally dark-colored Irish Moss would be rejected, so commer-

cial gatherers are forced to sun-bleach their product before it is sal-
able. Let me assure you that the dark-colored, fresh, or shade-dried
Irish Moss is better for almost any purpose than the blanched product that is on the market.

Freshly gathered Irish Moss is much tougher than either Laver
or Dulse, and unlike those two plants, it is not tenderized by drying.
As Irish Moss dries, it becomes tougher and harder, until it is almost
horny in texture. It is simply not suitable for eating raw, either fresh
or dried. Although Irish Moss is usually used only to make a gelatinous extract, the whole plant can be eaten, for, strangely enough,
this extremely tough plant becomes perfectly tender on being boiled
for only a few minutes.

After gathering a supply of Irish Moss, my first experiment was
in making blancmange. For this purpose, the fresh weed must be
thoroughly washed in several fresh waters, or the finished product
will taste fishy. Take as much fresh Irish Moss as can be packed into
1 cup, and tie it in a cheesecloth bag. Put this bag and 2 quarts of
milk in the top part of a double boiler, and cook over boiling water
for about 30 minutes. Stir occasionally, and press the bag against the
side of the pan with a spoon, to extract as much gelatinous material
as possible. If the milk is not stirred, the moss extract will tend to
coagulate around the bag. After the Irish Moss has steeped in the
hot milk for 30 minutes, give the bag a final squeeze and remove it,
discarding the spent seaweed. Add ¾ cup sugar and a pinch of salt
to the milk, and allow partly to cool. When it begins to thicken, add
any fruit or flavor desired, pour into molds, and chill in the refrigerator until it sets. I did my experimenting in Maine during the wild
berry season, so at various times I filled mine with freshly picked
blueberries, raspberries, shadberries, or all three together. Delicious.

Chopped Irish Moss can be added to nearly any soup or stew with
the other vegetables used, and it will serve the same purpose as okra
or the gumbo filé that is made of dried sassafras leaves, that is, it
will add body to the soup and keep it from being watery and uninteresting. If Irish Moss is covered with water and boiled for 30
minutes, it will make a soup that is almost tasteless but very nourishing, and if this soup is allowed to cool it will be a nourishing jelly.
I cannot understand why this food is not recommended in the various manuals on survival, unless it is simply because no one ever
thought of trying it. Other seaside foods are recommended, and cer-

tainly Irish Moss is abundant and easily available on Northern shores. I once made an entire meal on plain Irish Moss jelly, and while it was not delicious, it wasn't at all repugnant, and it slid down my throat so easily that it actually was pleasant. It satisfied my hunger and seemed to give me energy, and it produced no unpleasant stomach or bowel symptoms—rather the opposite—for it sat very well in my gastrointestinal tract, which is more than I can say for some of the other wild foods recommended in the survival manuals. Rock lichens, reindeer moss, and Iceland moss have all been recommended as emergency foods, cooked in this same way, but when I tried those three dry-land plants, I found them all both bitter and purgative, and they made me weaker rather than stronger.

Once my interest was aroused in seaweeds as emergency foods, I experimented further. I found that Dulse, Laver, and even the Edible Kelp described below could be cooked in this same way and eaten either hot as soup or cold as a nourishing but rather tasteless jelly. The resemblance of this jelly to aspic salad led me further. One day I gathered a handful each of Irish Moss, Laver, and Dulse, covered them well with water and set them to boil for 30 minutes. When this soup was cool but still not set, I stirred into it some chopped Orach, some Glasswort, and a very small amount of chopped Sea Rocket, and poured it all into a mold. It made a delicious aspic salad, and with all those mineral- and vitamin-rich plants it must have been about the most healthful salad ever made.

Recently, I made a "civilized" aspic salad using Irish Moss. For this I poured a 24-ounce can of a commercial mixture of eight vegetable juices into the top of a double boiler and added a cup of Irish Moss tied in a cheesecloth bag. After steeping it over boiling water for 30 minutes, I gave the moss bag a final squeeze, then added 2 tablespoons chopped onion, ¼ cup finely chopped celery, and 1 tablespoon finely chopped parsley to the juice. This was poured into a mold and allowed to set in the refrigerator. It was very pleasant, and even seemed to be an improvement over aspic made with ordinary gelatin.

The final edible seaweed that I collected from a rocky ledge in my low tide pool was Edible Kelp, *Alaria esculenta*. Please note that even the taxonomists have recognized this kelp as an edible plant, for the specific name, *esculenta*, means "edible." It is a typical Kelp, with a short cylindrical stem and a single main frond that is wavy or ruffled, 3 to 6 inches wide and from 1 to 10 feet long. This frond is thin, olive green or olive brown, and has a conspicuous midrib running from base to apex. At the base of this main frond are a number of ribless, secondary, lateral fronds that are borne on short, thin stalks that spring from each side of the main stem. These secondary fronds are tongue-shaped, from 1 to 3 inches broad, and from 3 to 10 inches long. Edible Kelp can be distinguished from other Kelps by these secondary lateral fronds and by the midrib in the main frond, for the only other Kelp that has a midrib in the frond is the Sea Colander, which has numerous round holes or perforations in the frond, not found in our Edible Kelp.

In my experiments I did not find the olive membrane of the main frond at all palatable, but when this was removed the thick midrib made a pleasant nibble, having a sweet taste. When sliced very thin, crosswise, this midrib made a pleasant addition to a tossed salad of wild seaside plants, tending to tame some of the wilder and stronger ingredients. The small, ribless lateral fronds also proved to be palatable, even when freshly gathered, and the flavor improved on partial drying. I gathered quite a quantity of these and dried them thoroughly. These completely dried fronds were somewhat stiffer than dried Laver, so I dipped them, a few at a time, in boiling water, leaving them immersed only a few seconds, to soften them. Then I carefully spread the fronds and stacked them one on another on a chopping board, then placed another chopping board over the stack, and squeezed the seaweed between them until it was tightly pressed and all excess water had run out. These seaweed cakes were placed in an oven turned as low as possible, and, to allow moisture to escape, I propped the oven door open slightly. The idea was not to cook them but to dry them. When they were thoroughly dried I cooled them, then shredded them with a sharp knife. The result was as good *kombu* as any that ever came from a Japanese factory. Maybe I'm prejudiced, but to me it seemed even better than Japanese *kombu*. I suspect that *Alaria esculenta* is a better species of Kelp than the Japanese use to make this food. I packed the shredded *kombu* loosely in fruit jars, set them in the oven until they were warm through, then sealed them; it kept perfectly until I had used it all up.

Kombu is used as a seasoning, as a vegetable, and as the chief ingredient in an "instant" soup. As a seasoning, rub shredded *kombu* between the hands to powder it, then sift out the larger pieces. Put the powdered product in a salt shaker, label it plainly, and put it on the table. It is excellent in all soups, stews, broths, and noodle dishes, and many like to add *kombu* to vegetables and meats. For an unusual vegetable dish, boil 1 cup shredded *kombu* with 2 cups water. It will take up most of the water and will be tender in 10 minutes of cooking. Drain any excess water and season with soy sauce, ½ teaspoon sugar, and 1 teaspoon monosodium glutamate. For instant soup, put 1 tablespoon shredded *kombu* in a cup and pour the cup full of boiling water. Season with soy sauce and monosodium glutamate. Like all new foods, these Kelp products take a bit of getting used to for the average Occidental, but they are well worth learning to like.

Why bother to learn to like seaweeds when there are plenty of other foods available that we already like? Health is one reason. The seaweeds are far richer in organic iodine and potassium than any other vegetable source, and they contain significant quantities of many other minerals the body needs for glowing health. Ever since Dr. Jarvis wrote his excellent book on *Vermont Folk Medicine,* which recommended Kelp, the sales of this seaweed have been zooming. But this is powdered Kelp in capsules, and therefore medicine. I would much rather *eat* than *take* medicine and find it far pleasanter to get my minerals in delicious vegetable and soup dishes than to take countless capsules. As Hippocrates, the father of medicine, said, "Leave your drugs in the chemist's pot if you cannot cure the patient with food."

27. Seaside Vegetables and Salads

IT seems strange that some plants actually prefer the inhospitable environment of rocky or gravelly surf-beaten shores with sterile, saline sand or soil, but such is the case. And these plants are not the tough customers you would expect them to be, but many of them are tender, succulent, and tasty. Around most of North America, the upper beaches of saltwater shores are veritable vegetable gardens for those who can glean them with skilled hands and a knowing eye. From early spring until late autumn, these shores will always furnish a vegetable dish to go with the seafood that the nearby ocean or intertidal shore is ready to contribute to all who approach it with an humble spirit and a readiness to let it teach you its secrets.

ORACH OR SEASIDE LAMB'S-QUARTERS
Atriplex, several species

Orach looks enough like the common Pigweed to be mistaken for the same plant, but since it differs in its habitat and the way it bears its flowers and seeds, botanists have very rightly put it into a separate genus. The plant is hard to describe, as it varies so greatly, in some places being but a single stem sprawled over the rounded stones of the upper beach, and in others growing rankly upright and

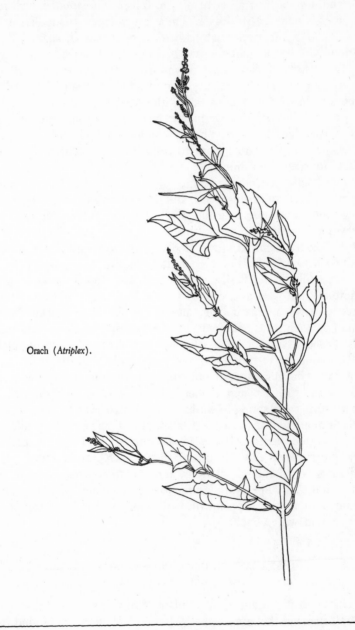

Orach (*Atriplex*).

sometimes reaching a height of 4 to 6 feet. The amateur will recognize the plant by its leaves. These are roughly triangular, 1 to 3 inches long and somewhat halberd-shaped, that is, shaped like an arrowhead, except that the two barbs on the lower edges are turned outward instead of pointing downward. These leaves are dark green in color and somewhat scurfy or mealy in appearance, especially on the lower surfaces. It grows along the seashore, just above the high-water line, from Labrador to Virginia, and in many districts is so abundant that it would be hard to find a hundred yards of shoreline without enough of this plant to furnish an excellent salad or vegetable for the camp dinner.

Orach, like the familiar Pigweed, is closely related to our garden spinach, beet, and Swiss chard, and used either as a salad or a cooked vegetable, it is more than the equal of any of them. The dark-green leaves are tender, succulent, and tasty, and grow so close to the sea that they are slightly impregnated with sea salt with all its healthful minerals. Orach needs no preparation whatever to be good food, and I often nibble the fleshy, ready-salted leaves as I walk along the shore. When used as a cooked vegetable, Orach, unlike many wild greens, needs no parboiling, and it can be short-cooked like the finest spinach, thus conserving all the healthful vitamins and minerals; it doesn't even require salting—just boil the juicy leaves in very little water for 10 minutes, butter generously, and serve.

Orach continues to put out new growth throughout the season, so you can gather young, tender leaves in just the right stage for the best eating from spring to winter. Tossed with your favorite dressing, Orach alone will make a good salad, and it mixes wonderfully with other seaside plants or domestic salad materials. There are probably few green plants more healthful than Orach, for besides generous offerings of chlorophyl, vitamins, and sea-salt minerals, it is an iron-rich plant that will do wonders for your "tired blood." For a vegetable tonic, a free diet supplement, and some really good eating, get acquainted with Orach.

GLASSWORT, CHICKEN CLAWS, OR SAMPHIRE
Salicornia, several species

Glasswort is a beautiful little plant that grows on clay shores and in salt marshes that are lapped by the highest tides. It is found from

Massachusetts to Florida, and around the Gulf Coast, wherever there is heavy soil and tidal water. It, too, is a relative of our domestic beet, spinach, and Swiss chard, but here the relationship is apparent only to the expert.

There are three species of this plant on our shores, but they are all edible and look much alike, so they can be described together. In spring and early summer, a patch of Glasswort may be no more than a group of little green spikes, 2 to 3 inches high and about ¼ inch in diameter, of a clear, translucent green. As it grows it becomes conspicuously jointed, with opposite branches at the joints, gradually assuming the shape of a tiny, sparse, leafless Christmas tree. It appears to be completely leafless, and it takes a trained botanist to recognize the tiny scales at the joints as degenerate leaves. The flowers and seeds are even more inconspicuous, being hidden under the almost invisible leaves. Just look for the cylindrical, ¼-inch-thick, branched, leafless stems, rising a few inches above clay shores that are submerged by the highest tides. In autumn it turns a brilliant red, beginning at the base, but the top, tender growth remains edible from the time it appears in the spring until the first hard freeze of winter.

When you find a patch of these little cylindrical stems, pinch off only the upper, tender part. Taste one right on the beach. The taste is clean, juicy, and slightly salty. It makes a pleasant nibble and is well worth carrying to camp or home for a salad, garnish, cooked vegetable, or pickle.

Glasswort makes a beautiful green salad. Just toss a handful of the whole plants with your favorite spiced French dressing. The clear emerald color and interesting shape are very attractive, and the young stems are so tender that it is easy to cut them into bite sizes with a salad fork. Glasswort can also be sliced or chopped and mixed with almost any salad green, wild or domestic, and it will contribute a clean, sprightly flavor and a salty tang that will be appreciated.

Use bright green sprigs of Glasswort to garnish a planked fish, a dish of fried clams, soft-shell crabs, or almost any other seafood. Glasswort is as pretty a plant as ever came to the table, and a great deal better than many domestic plants that we commonly eat and relish.

Glasswort makes a passable cooked vegetable boiled for just 10 minutes and generously buttered. However, when cooked it is not as good as some of the other seaside plants named in this chapter, so I would advise that the Glasswort you find be reserved for salads and garnishes, and that you use Orach, Sea Blite, or Goosetongue as cooked vegetables.

In England, quite a different fleshy seaside plant, *Crithmum maritimum*, was called "Samphire" and was commonly used to make pickles. English settlers along our coasts soon discovered that our Glasswort was an acceptable substitute, if not actually better than the real Samphire, so they transferred the name and the lore to this new plant. Therefore, in many districts today you will find Glasswort called Samphire. I tried, several times, to make "Samphire pickles" of Glasswort, using ancient recipes that I found in old cookbooks, but I found all of them too much work, or else they produced a pickle too soft and stringy for my taste, so I devised my own method of making these smart-looking pickles. Wash freshly gathered Glasswort and pack into pint jars with the stems straight and vertical. Make a pickle of 1 quart vinegar, ½ cup sugar, 3 tablespoons mixed pickling spices, one sliced onion, and six dried Bayberry leaves. Boil together 10 minutes, then pour boiling hot over the Glasswort until the jars are level full. Seal and store for three weeks before broaching. This makes a beautiful pack, and the pickles taste as good as they look.

SEA BLITE
Suaeda, five species

Sea Blite is still another member of the family that gives us Orach, Glasswort, and several domestic vegetables. It is an humble, inconspicuous member of the family but it, too, furnishes very good cooked greens. It grows on upper beaches from Labrador to Florida, and on around the Gulf Coast. The species are so much alike that the original edition of Gray's *Manual of Botany* lumped them all into one species, *Suaeda maritima*, but subsequently, more sophisticated taxonomists have found sufficient differences to justify dividing the genus into five species, which need not bother us, as they are all equally edible.

Sea Blite doesn't *look* edible. The plants grow singly, scattered over

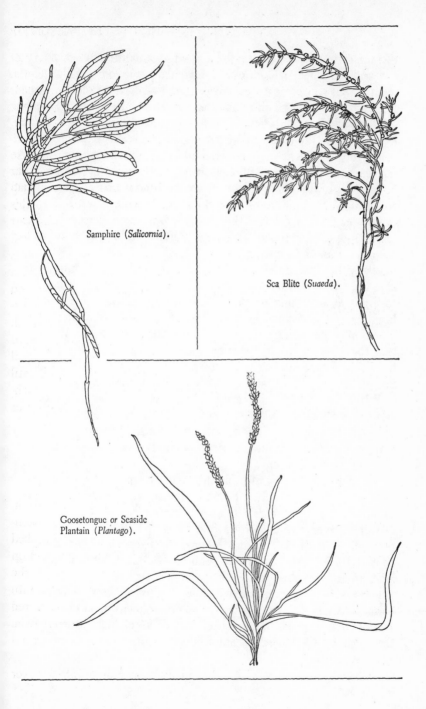

Samphire (*Salicornia*).

Sea Blite (*Suaeda*).

Goosetongue *or* Seaside
Plantain (*Plantago*).

the upper beach just above the line of tidal-borne sea wrack. It is low and branched, seldom over a foot high, and some of the plants seem to have given up the struggle and lie prostrate on the beach. Its leaves, instead of being broad and green as in the related Orach, are tiny, thick, and narrow, almost cylindrical, and not over 1 inch long. In color the whole plant has a gray-green, dryish look but it is really very succulent and tender. Unlike many wild greens, you do not have to gather Sea Blite at just exactly the right stage; it is edible over quite a long season, from the time it first appears in the spring until the middle of August, in the Northern states.

The leaves and calyxes, the inconspicuous green flowers, and the tender upper stems are all edible. These tender parts are easily stripped from the growing plant.

Like other seaside plants, Sea Blite is ready-salted, but in this case the salting has been somewhat overdone. Usually it is necessary to drain at least one water, to reduce the salinity to a palatable level. Place the cleaned leaves, tender stems, and even young flower bracts in a stewpan and pour boiling water over them. As you do this the unattractive gray color will disappear and the dish will become a beautiful bright green. Boil for 5 minutes, then drain. Serve it chopped and buttered, as you would spinach.

While Sea Blite is not very good raw, it does make an excellent salad if cooked first. Chop 2 cups of boiled and drained Sea Blite in small bits with a pair of kitchen shears, then stir in ¼ cup mayonnaise. Smooth into a pretty mound, and garnish with thin slices of hard-boiled egg.

Another good way is to cook, drain, and chop 4 cups of Sea Blite, then stir in 1 can of either cream of mushroom, or cream of celery, soup, undiluted. Heat to a simmer, then serve.

Still another way is to fry a few slices of bacon until crisp, then pour off most of the fat and remove the bacon. Put the chopped, cooked Blite in the bacon pan and dry it out a bit over medium heat, then serve with the bacon crumbled over it.

If you are inventive, you will find dozens of ways of using this versatile vegetable. It mixes well with other greens, wild or domestic, and deserves a place in any mixed bag of vegetables gathered from the seashore. Get to know Sea Blite.

GOOSETONGUE OR SEASIDE PLANTAIN
Plantago oliganthos and *P. juncoides*

This useful little plant is actually a relative of that most pestiferous of lawn weeds, the common dooryard plantain, but Goosetongue is the lovable member of the family. It is a perennial, growing in clumps along clay shores and even in the cracks and depressions in the rocks of sea bluffs where nothing else will grow. The slender, fleshy leaves are gray-green, 4 to 6 inches long and about half an inch wide, and resemble succulent, brittle grass-blades. It carries spikes of drab, greenish flowers on slender stalks 4 to 8 inches high. It is found in great abundance on both the Atlantic and the Pacific coasts, in the East from Hudson Bay to New Jersey and in the West from far up in Alaska to California.

Despite its abundance and wide distribution, few people seem to know that Goosetongue is both an excellent salad plant and a palatable vegetable when cooked. The crisp leaves can be snapped like green beans, and when boiled for 20 minutes in very little water and served with plenty of butter, they make as nutritious and tasty a vegetable as could be desired.

The tender, brittle leaves can be broken or cut, dressed with a little oil and vinegar, and they make a crisp, attractive salad, or they can be combined with other salad materials and will contribute a salty crispiness that nearly everyone likes.

While on a recent camping trip in Maine, my wife and I decided one afternoon to scrounge the makings for a big clam chowder for our dinner that night. We parked by the end of a bridge where mounds of clam shells showed that this place had often been used by commercial clammers to shuck their catch. A little exploring revealed that we had found not only clams but an entire food complex.

Out on the flat, numerous squirts and holes betrayed the presence of an abundance of clams that were easily obtained with a clam fork. There were large rocks scattered about on the flat, crawling with fat Periwinkles and Dog Whelks, and lined around their bases with Blue Mussels, large ones. In a few minutes we had all the seafood we could possibly eat, and some of the Dog Whelks to use as bait in the fishing we planned for the next day.

Among the beach stones, just above high-water mark, we found

Orach, Glasswort, and Goosetongue, all growing together in great profusion. Not six feet away on the landward side were thick clumps of Sour Sorrel. We picked a bit of each and a little more of the Goosetongue, for both of us are very fond of it either raw or cooked.

After leaving this wild vegetable garden, we started inland toward the road and came on a regular briar patch of wild raspberries hanging red and ripe. We drank the coffee from our vacuum bottle and filled it with these delicious berries. I saw a red glint nearby that on investigation proved to be shadberries, hanging in thick, ripe clusters and big as the end of your finger, and at the bases of the shadbushes grew low, early blueberries just right for picking. We had used all our containers, so I filled my hat with these generous offerings, and we arrived back at the car, having been gone only an hour, laden with a grand assortment of seafood, wild vegetables, and wild berries, far more than enough for a delicious dinner for which the food had almost miraculously appeared as a by-product of our recreation.

Had we accidentally stumbled onto a unique paradise? Not at all. This foraging performance could be repeated, with variations in species, of course, in thousands of places along the saltwater shores that border this grand continent of North America.

SCURVY GRASS: RELISH AND SALAD
Cochlearia, three species

In the days when wooden ships were sailed by iron men, the greatest hazard of the sailor's life was neither storms nor calms, nor was it shipwrecks or pirates; it was a disease called scurvy. Before the development of the canning industry and modern refrigeration, seamen on long voyages were forced to live on salted and dried foods, a diet that lacked certain elements we now know to be essential in human nutrition. Scurvy, a vitamin-deficiency disease, was an accepted occupational hazard of the sailor's life. On long voyages it was expected that some of the crew would die of this malady, and sometimes whole crews became so afflicted that there were not enough able men left to handle the ship.

It was not only sailors who suffered from scurvy. Early colonists on strange shores were often defeated by this same disease. Knowing the native flora would probably be unfamiliar, these early settlers often brought with them salted or dried food to last until they could clear the land and farm. When they tried to live on these vitaminless rations many of them became scurvy victims. The saddest part of all this early suffering was that the means of preventing or curing scurvy was usually growing all around, had the colonists only known which greens to gather.

Long before anyone had ever heard the word "vitamin," people noticed that when scurvy-ridden sailors obtained fresh fruit and vegetables they quickly recovered. It is a miracle, the way a deficiency often creates a craving, and like many vitamin-starved sailors, the colonists began to crave fresh vegetables. In their extremity they started experimenting with wild vegetation, and they can be credited with discovering many plants that we cultivate today.

Early voyagers to the northeastern shores of our continent found a plant growing along the rocky beaches and on the sea cliffs that was so efficacious in preventing and curing this dread malady that they called it "Scurvy Grass." It was not only healthful, but as palatable as the finest cress, and could be obtained throughout the ice-free season of the year. Scurvy Grass is not really a grass, but is a member of the mustard family, and it has the familiar, mild pungency and piquant flavor that makes so many members of this family favorite vegetables, for Scurvy Grass is closely related to our cress, cabbage, cauliflower, broccoli, radish, and garden mustard. It is a small plant with roundish leaf blades that grow in a fan-shaped angle on the slender main stem, forming a circular rosette about 6 inches across on an average. The flower stalk rises from the center and is only a few inches high, bearing white flowers that, like all the mustard family, have 4 petals and 6 stamens. The fleshy and almost veinless leaves and even the tender leaf stems of Scurvy Grass furnish one of the tastiest of all wild greens, even rivaling the swanky watercress in palatability.

Scurvy Grass is abundant around the shores and islands of the Gulf of Saint Lawrence, Newfoundland, and outer Labrador, and as much as a camper could conceivably need can usually be found along the shores of Nova Scotia and on the islands of the Maine coast. All of this wonderful area is now accessible to campers, and there is every reason for going there. The fishing is unbelievably good, and the clams, periwinkles, urchins, mussels, and many other strange and delicious seafoods are to be found all along these shores.

The lobsters and crabs are wonderful, and on the shores one finds wild strawberries, raspberries, shadberries, blueberries, waterberries, and mountain cranberries in their various seasons. Not the least of the attractions of this northern area is the abundance of crisp Scurvy Grass, furnishing one of the most delicious of wild salads and relishes.

Use Scurvy Grass as a salad or an edible garnish. A very good sandwich can be made of nothing but bread, butter, and Scurvy Grass, or the Scurvy Grass can be added to most meat sandwiches and it will improve them in flavor as well as in wholesomeness. Short-cook Scurvy Grass as you would the tenderest spinach, and it will furnish a very palatable green vegetable. Use it to stuff some of the Northern Flounder you can catch along these shores, and you will have a gourmet dish that only you and nature have created.

SEA ROCKET: POTHERB AND SALAD
Cakile edentula

Another seaside plant of the mustard family is Sea Rocket, and on both Atlantic and Pacific shores it can add zest to a sandwich, enrich a tossed salad, or furnish a vegetable dish. It is a branched, fleshy annual from 1 to 2 feet high found in abundance on sandy and gravelly shores from New Jersey to Labrador and in many places along the Pacific from California to British Columbia. The upper stems are tender and succulent; the leaves—about 2 inches long, fleshy and wedge-shaped, and attached to the stem by the narrow end—are wavy-margined, with a flavor reminiscent of mustard greens and horseradish with a hint of garlic. The plant can be certainly identified by the flowers, which are borne on the upper stems and are lavender, only about ¼ inch across; like all flowers in the mustard family, they are cross-shaped, with 4 petals in two opposite pairs, and there are 6 stamens. The flowers are followed by distinctive plump, green seedpods that are two-jointed, the lower joint like a round ball about ¼ inch in diameter and the upper joint sharply beaked, the whole seedpod about 1 inch long.

The tender stems, the fleshy leaves, the green seedpods, and the flower heads are all edible. The leaves and succulent seed capsules are excellent cut into a tossed salad with milder greens like Samphire, Orach, or Goosetongue. The tenderest leaves will add real zest to a sandwich, the flavor being reminiscent of mustard greeens and horseradish. A liverwurst sandwich sounds anything but epicurean, but when the liverwurst is on hot toasted and buttered rye bread and dressed with a layer of Sea Rocket leaves, this proletarian food suddenly becomes aristocratic.

The flower buds, green seedpods, tender stems, and leaves can all be cooked as vegetables, either together or separately. One would think that a plant so sprightly and pungent when raw would make strong-tasting greens when cooked, but not so. As it cooks it gives off a turniplike odor, but the strong, horseradish-like flavor must be very volatile, for the cooked greens are mild and pleasant. Eat them liberally buttered.

I like to walk along the shore, tailoring a vegetable dish as I explore, adding a bit of Orach, some more of Goosetongue, a handful of Sea Rocket, and whatever other wild vegetables I find growing there, mentally combining flavors, and constructing a dish that has just enough of each to suit my fancy. One can do the same in building a wild seaside salad.

Once, in northern Maine, we made a jell of Irish Moss, then combined Goosetongue, Orach, Sea Rocket and Sour Sorrel to make an aspic salad in individual molds. When it was set, we inserted a branched sprig of translucent Glasswort—like a little emerald-green Christmas tree—in the top of each. It was pretty enough to serve at the fanciest kind of dinner party, good enough to set before the most fastidious epicure, and wholesome enough to delight the most enthusiastic health-food faddist. Every ingredient, even to the sea salt that gave it savor, had been freely given by the sea and its fringing shore. Wild food need not be crude or primitive fare. It can contribute to gracious living if given the preparation and attention that its fine flavors really deserve, and then presented with style and imagination.

ROSEROOT: A PERFUMED SALAD PLANT
Sedum rosea

This is another interesting little plant that has earned the title "Scurvy Grass," and that is what it is commonly called in the northern part of its range. Found on ledges, banks, and sea cliffs from far up in the Arctic to as far south as the Delaware River, it is abundant in the northern part of its range and becomes rarer southward. To sailors, fishermen, explorers, and natives of the far north, this is an important salad plant in a region where green vegetables are very scarce. To the seaside camper and tourist of Maine, Nova Scotia, Newfoundland, and other interesting northern regions, it can furnish a dainty salad to soften the harsh seaside fare.

It is a very distinctive and easily recognized little plant standing only 6 to 10 inches high, the stem crowded and usually hidden with fleshy, whitish-green leaves tinged with pink. The individual leaves are about 1 inch long, oblong in shape with toothed margins. The flowers are yellow, sometimes turning purplish later, and are borne in a dense cluster atop the stem, followed by several-pronged seed capsules that are purple or red in color. The root is perennial, rough, and large for so small a plant, and when bruised it gives off a fragrance like rose-scented toilet water.

The tender stems and succulent leaves are mild and juicy, with a clean, refreshing taste. Often found growing in company with that other "Scurvy Grass," *Cochlearia*, the two can be chopped together to make a delicious arctic salad that is fairly loaded with antiscorbutic vitamins and healthful minerals. Any time before the seed capsules form, one can gather the stems with the leaves intact and cook them like asparagus. With plenty of butter and a little black pepper, they make a very tasty vegetable.

Unlike many of the plants mentioned in this section, Roseroot is not akin to any of our garden vegetables, but it is a close relative of the common Orpine, or Live-Forever, *Sedum purpuream*, which is sometimes called Aaron's-Rod or Frog-Plant. This latter *Sedum* somewhat resembles its little arctic cousin, except that Live-Forever is a much stouter and ranker-growing plant. It, too, has very succulent, fleshy leaves, oblong with toothed edges, crowded spirally onto a stout stem 1 to 2 feet high. In midsummer the plant is crowned with a broad, rounded, and very attractive cluster of small red flowers. It was brought to America from Europe as a garden ornamental, but long ago escaped to the wild and is now an almost too-familiar plant of damp fields, roadsides, and rich, rocky banks, becoming almost a weed from Maryland to eastern Quebec.

Despite its abundance and availability, few people seem to know that Live-Forever is also a wild-food plant of no mean order. Like its northern seaside relative, it has stems and leaves that are tender,

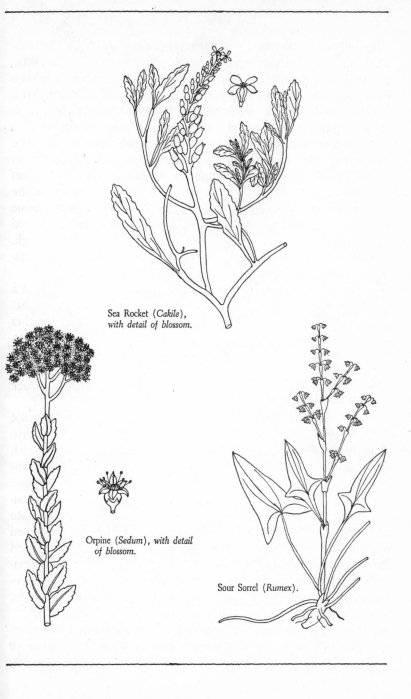

Sea Rocket (*Cakile*),
with detail of blossom.

Orpine (*Sedum*), with detail
of blossom.

Sour Sorrel (*Rumex*).

mild, and tasty. For use as a salad, the plant should be gathered when very young, and then the stems and leaves can be sliced together and served alone or mixed with other salad plants, wild or domestic. Any time before it flowers, the stems, either with or without the crowded leaves, can be cooked and served in the same manner as asparagus. In addition, the large root of this remarkable plant bears numerous white, crisp tubers the size and shape of your little finger that can be sliced into a salad, eaten as a crisp, radishlike appetizer, or made into a pickle. Take only the small, light-colored, recently grown tubers, as the older ones tend to be tough and stringy. These are produced in great abundance through the spring and early summer, and then another crop is put on in the late autumn.

To pickle wild Live-Forever tubers, wash the crisp young ones and pack into jars. Cover with a pickle made in the proportions of 2 cups vinegar to 1 tablespoon mixed pickling spices and 2 teaspoons salt. Boil together 10 minutes, then pour over the tubers boiling hot. Seal immediately with dome lids, and store in a dark closet for at least three weeks before using.

Although Live-Forever and the two ferns mentioned in the following section are not strictly seaside plants, they are abundantly available to the seaside camper who will take the trouble to walk a bit inland on his foraging excursions, and since I have not given directions for their use in any of my other works on wild food, I think it proper to discuss them here.

HOW TO EAT A FERN
Peteridium and *Pteretis* species

I love the early spring. In fact I love it so well that I sometimes experience two springs in the same year. In Pennsylvania, that great meeting place of northern and southern flora, where I live, the early springs are magnificent. After a sedentary winter spent largely before my typewriter, I fairly revel in the awakening vegetation. Then, when the first spring flowers begin to fade near my home, I pack my camping gear and head for the rocky coasts of the New England states and experience it all over again.

There are many advantages to early spring camping in New England. The summer people have not begun to arrive, and the local residents take you to their hearts with a hospitality that is lacking once the hordes of tourists begin to pour in. There are no bugs, mosquitoes, or ants to annoy you, and in clear weather you can sleep out under the stars. The nights are nippy, and you will need extra blankets and warm sleeping bags, but the cool evenings only make a campfire more enjoyable. If you prefer more luxury, cabins, cottages, and motels can be had at a fraction of what they will cost during the height of the tourist season.

This is the time when smelt and alewives run so thickly up their spawning streams that a few dips of the net will yield enough for dinner. At this time, too, the tremendous tides of that part of the world run farther out than they will again until the following autumn, exposing practically untouched flats, alive with Steamer Clams and Razor Shells. Many other seafoods are abundant and at their best in early spring. Chicory and dandelions are just right for salads and cooked greens. Not the least of the pleasures of early spring in New England is the enjoyment of fern fiddleheads, one of the greatest delicacies among wild vegetables.

I dislike the way the literature on this subject refers to fern fiddleheads as "a substitute for asparagus." Fern fiddleheads are perfectly good vegetables in their own right, with a characteristic flavor and texture of their own, and they are no more a substitute for asparagus than asparagus is a substitute for fern fiddleheads.

The edible part of the fern is the young frond while it is still tightly coiled in early spring. It is ready to eat long before the first asparagus. The spirally coiled young fronds are called "fiddleheads" from their resemblance to the tuning head of a violin, or "crosiers" from the curved ceremonial staff carried by a bishop. Many true ferns produce edible crosiers, but the best and at the same time the most plentiful and easily found fiddleheads are produced by the Brackens and Ostrich Ferns.

Bracken (*Pteridium*), also called Pasture Brake, is probably our commonest and best-known fern, for it grows from Newfoundland to Alaska and south to Mexico. Found in open, dry woods, clearings, pastures, and especially where forest fires have run, this is a coarse fern, the old fronds three-forked and attaining a height of several feet under favorable circumstances. The species are much alike in appearance and all of them have edible crosiers, so we'll let the taxonomists argue about how many species of *Pteridium* grow in North America, and we'll enjoy the crosiers of any that we find, even if we don't know one of its Latin names.

The way to locate a supply of the young fiddleheads is to look for the old last-year's fronds which persist over the winter. Around the bases of these old fronds, you will find the little uncoiling fronds of this year's crop, long before the first vegetables from the garden are ready to eat. These little ferns are solitary or scattered, about ½ inch thick at the base, round and covered with a rusty, cottony felt, the little fiddleheads at the summit distinctly three-forked. Take only the part that is tender enough to snap easily between your fingers. Pull each coiled frond through your hand to remove its natal wool, and they are ready to eat, raw or cooked.

The Ostrich Fern, *Pteretis nodulosa*, is found from coast to coast in the north, but doesn't extend as far south as the Brackens, although I have collected good Ostrich Fern crosiers in Maryland and Virginia. These delicious little fiddleheads spring from a perennial crown and form dense, vase-like clusters, so that a very few crowns will furnish a large dish of vegetables. Those bearing crowns are located by looking for the dried, persistent, last-year's fruiting fronds, resembling thick, dark-brown plumes several feet high. Look for these in alluvial soil along rich stream-beds, but in the north, where this fern abounds, you will always see all you could possibly want from the roads as you drive about in your car. Later these fiddleheads develop into great clumps of unbranched, feathery fronds, often standing waist-high and in ideal circumstances reaching a height of 9 feet. These old fronds are tough and inedible, but summer, when they are conspicuous, is a good time to locate a fruitful spot to come plundering next spring. Take only the tender little fronds with tightly coiled tops that stand 6 to 10 inches high. Do not pick all the young fronds from any one of the crowns. Take only the best and leave some to develop, and your foraging will not harm this beautiful plant. The fiddleheads you gather will be more or less covered with brownish scales that should be removed before cooking if you want to avoid a taste of tannin, but they are loosely fastened and easily rub off.

Whether you use Bracken or Ostrich Fern, you will find few

vegetables that can top a piece of toast with more grace and goodness than fern fiddleheads. They can be boiled like asparagus, sautéed in butter, served in a cream sauce, or, the best way of all, steamed and served on toast with a wine Hollandaise Sauce.

To steam ferns, place the cleaned and washed fiddleheads in a colander over boiling water for 30 minutes. To make the sauce, melt ¼ pound of butter in a saucepan, and while it is melting, beat 3 egg yolks until they are thick and lemon-colored. While still beating, add ½ teaspoon salt and a few drops of Tabasco sauce to the yolks, then beat in about ⅓ of the hot, melted butter. Add 4 teaspoons lemon juice and 4 teaspoons white wine or sherry, a teaspoonful at a time, alternating with the addition of the rest of the hot, melted butter, also added in small quantities at a time. Beat constantly until all has been added. Since this Hollandaise is never cooked, there is no danger of curdling or separation, and there is no excuse for failure if you follow the directions to keep beating and to add the ingredients very gradually. It can be made ahead and stored in the refrigerator, or even kept for that dinner you are planning for next week. It will stiffen up a bit under refrigeration, but will melt instantly when placed on hot vegetables or other food. To use it with ferns, salt the steamed fiddleheads and pile them while still hot on a hot piece of toast, then spoon a generous helping of the sauce over them.

Many people like Bracken fiddleheads raw in salads. Mix with other salad materials, and use an oil dressing. Both kinds of fiddleheads can be stored in the freezer for out-of-season use. Clean and wash them, then dip in boiling water for about 2 minutes, cool as rapidly as possible, then package and place in the quick-freeze section of your freezer. These frozen fiddleheads should be short-cooked when used (cook not over 10 minutes) and then served exactly as you handle fresh ones.

Fern fiddleheads have long been greatly appreciated in Japan and New Zealand, but few Americans have ever tasted them. In Japan a starch is refined from the roots, but this destroys the plant, so drastic laws had to be passed to preserve the ferns. Judiciously gathering the uncoiling fronds does not hurt the ferns, or I would not recommend them as food, for I think these ferns some of the loveliest vegetation in our native flora, and the parent plants should be preserved at all costs.

PRICKLY PEAR
Opuntia, several species

The Prickly Pear is the fruit of a cactus, and reaches its greatest development in the arid Southwest, but the Eastern Prickly Pear, *Opuntia vulgaris,* grows in sand and rocks from Massachusetts to Florida and Alabama, generally not far from the sea, and so merits inclusion here. It is a thickened, jointed, branching cactus, usually prostrate, but sometimes ascending for a joint or two. The joints, often called the "leaves," are oval and flat, from 2 to 6 inches long, bearing a few solitary, sharp-pointed spines about an inch long. Occasionally one finds a plant with no spines at all. The real leaves of this plant are tiny, awl-shaped protuberances that lie flat against the joint and generally drop early. In spring and early summer, this cactus bears beautiful yellow flowers about 2 inches across, with 8 or 10 petals and numerous stamens. These are followed by pear-shaped fruits from 1 to 2 inches long that become red, juicy, and edible when ripe. You cannot fail to recognize this plant when you come on it, for there are no other plants that remotely resemble it in the area where the Eastern Prickly Pear grows. Once you are acquainted with one species of *Opuntia,* others in the genus are easily recognized, and all of them that have plump, smooth fruits are edible when ripe.

Prickly Pears have a pleasant taste, and may be eaten either raw or stewed. Unfortunately, the surface of the fruit is covered with tufts of lightly fastened and extremely irritating bristly stickers that all too easily get into fingers or lips if the fruit is not skillfully handled. The Mexicans, Indians, and boys of all races where this fruit abounds have learned to overcome this difficulty, and the fruit is highly prized. I have seen de-bristled Prickly Pears of several species being sold in the markets of New York and Philadelphia, at famine prices.

It was in Texas and New Mexico that I first learned the technique of handling Prickly Pears, and it works just as well in any part of the country where you find them. The Indians taught me to take a handful of grass or weed and brush the bristles from the fruit while it was still attached to the stem. Then the fruit was cut from the cactus, and as a final safety measure, each fruit was rubbed with a damp cloth. They are delicious eaten out of hand, and the small varieties can be eaten skin and all. The larger kinds are better if one slices off a bit of both the stem and blossom ends and peels the fruit; then it should be thoroughly chilled, when it will taste like watermelon, only more so.

I remember patches of Prickly Pear in New Mexico that had yellow fruit, sweet as sugar, and another patch that produced plump, almost round pears that were purple in color, with a flavor of fresh figs. I have eaten the very large, sweet, white pears of this family that grow in the highlands of the Hawaiian Islands and are in season for five months of the year. I have tasted surgary purple ones from Puerto Rico, and large red ones that were planted by the Carmelite monks in California. However, our lowly Eastern Prickly Pear is not the worst of these. It is small but flavorful, and in my opinion well worth the effort it takes to gather and prepare it. It is probably best eaten raw, but also makes a good stewed fruit and is so sweet that it requires very little sugar when cooked this way.

The Prickly Pears of all species also can be used as vegetables. The green fruits when gathered shortly after the blooms fall can be added to soups and stew and will impart an okra-like mucilaginous quality. The flat, leaflike stem joints can also be used. Only half-grown tender upper joints should be gathered for this purpose. These should be speared with a skewer and roasted over an open flame until all spines and bristles are singed away. Then peel them, and the soft inside can be either boiled or rolled in cornmeal and fried. The fried product will be more pleasing to most palates, for the boiled stems, while palatable, tend to be very mucilaginous and —let's face it—somewhat slimy, but this is a quality that one can learn not only to tolerate, but even to like. These boiled joints of *Opuntia* are a favorite vegetable in Mexico, and are often sold by street-side venders who cook them on the spot. If you are a purist about your wild food, substitute these peeled stems of Prickly Pear for the okra called for in the recipe for Crab Gumbo on page 46, and the dish will actually be improved.

THE BEACH PEA
Lathyrus maritimus

The Beach Pea immediately suggests itself as a source of wild food to all who see it, because its resemblance to garden peas is so obvious. From New Jersey northward, on the Atlantic, and from California to the Bering Sea, on the Pacific, it luxuriates on sea strands so sterile that even grass will not grow there. It is also found about the Great Lakes. Usually growing in patches, the individual plants are about a foot high with compound, pea-like leaves each bearing from 3 to 5 pairs of leaflets. The sweet-pea-like blooms are bright purple and quite pretty, and they are followed by heavy crops of small peas that can be gathered by the handful. Shelled, these peas closely resemble garden peas except for size, as the Beach Peas are much smaller.

Other writers on this subject have complained that Beach Peas are rather dry and tasteless, but they must have tried some that had already passed their prime. The finest of garden peas are also dry and tasteless if allowed to get too old before they are eaten. If one takes only the small, tender Beach Peas that are still bright green in color, they are delicious. I enjoyed them three times last summer, and my objection was not to their taste, which was very fine, but to the tediousness of shelling enough for a meal. Growing in clusters, they are very easy to gather, but shelling enough for a single serving is a formidable task taking more than an hour. If you intend to serve Beach Peas to several people, you had better draft them all for the shelling operation; then they will enjoy these delicious little peas even more, if only for the labor they have invested in them.

To cook Beach Peas, put them in enough boiling, salted water barely to cover and cook for 15 minutes. Drain, then to each serving add 1 tablespoon coffee cream and a little freshly ground black pepper. Serve while still hot, as they become uninteresting when cold.

STRAND WHEAT
Elymus arenarius

Ever since the eleventh century, Strand Wheat has been a staple grain in Iceland. Some of their earliest sagas tell of the ways in which this grain is prepared. To this day, there are those who, preferring a good taste to mere convenience, continue to gather the heads of Strand Wheat from the beaches and sea strands of Iceland. Why Strand Wheat has fallen into disuse in our own country, while wild rice has gone on to be acclaimed as a gourmet's delight, I'll never know, for the two grains, although not closely related, are much alike in use and flavor, and Strand Wheat is much the easier grain to harvest.

Strand Wheat, also called Wild Rye and Sea Lyme Grass in some sections, grows in clusters or patches, sometimes extensive enough to resemble fields of cultivated grain, on northern beaches and sea strands from Maine to Greenland in the east and from California to the Arctic on the Pacific coast. It is a coarse grass, about 3 feet high with whitish, flat leaves that become rolled in at the tip late in the season. The grain is borne in a spike or head, 3 to 8 inches long, at the top of the stalk. There are sometimes 30 to 40 grains in each head. In late August or early September, when the grain is ripe, it is harvested by pulling the heads from the stalks and dropping them into a basket. These heads are threshed by rubbing with the hands or by putting them on a smooth floor and walking over them with heavy shoes. This threshed grain is still enclosed in a husk or glume, just as is wild rice or domestic oats or barley.

Removing the husk from the grain is the big job in preparing Strand Wheat. The best way I have found to do this is first to thresh as much as will fit into all the shallow pans I can put in my oven at one time. Then I parch the grain in a very hot oven for about 1 hour, stirring frequently so it will parch evenly all over. These parched husks are very brittle and break off easily if the grain is pounded or rubbed between the hands; then the grain is cleaned by winnowing in a breeze. A tedious operation? Yes, it takes some time and work to get Strand Wheat ready for use, but if you figure the value of the grain at the prices you would have to pay for wild rice, you will find that you are making very good wages.

Other writers on wild foods have recommended Strand Wheat only as a breadstuff, but they have overlooked a good thing. The whole grain, cooked in the same manner as rice, is delicious with wild game or other meats, and is very good served with sugar and cream as a breakfast cereal. Thoroughly wash the grain in cold water, then put it into a heavy kettle and cover 1 inch deep with water. Add ¼ teaspoon salt for every cup of grain. Boil rapidly for 8 minutes, then reduce the heat, cover the kettle tightly, and cook slowly until all the water is absorbed, about 40 minutes. You will find this is more chewy than wild rice, and it has an unusual and delicious flavor all its own. As a breakfast cereal, it is even better if coarsely ground before cooking. I use an old-fashioned hand coffee grinder to make my Strand

Wheat grits. The only objection to these grits as a breakfast cereal is that the cook has to get up very early in order to cook them properly, as they are better if cooked for at least 1 hour. An old-fashioned fireless cooker, with the grain put on to cook the night before, would probably be the ideal way to cook this cereal.

When cleaned and ground fine in a grist mill, Strand Wheat can be used in the same ways as regular wheat flour. To improve the texture of the baked goods and make my supply of Strand Wheat flour go further, I usually mix it half-and-half with commercial flour, and the resulting pancakes, biscuits, and muffins have a rich and delicious flavor.

WOUNDWORT AND BUGLEWEEDS
The Tuber-Bearing Mints

These plants are found inland as well as near the sea, but it is only in open sand on the coastal plain that I have ever found the tubers abundant and available enough to use, so I always associate these two fine wild vegetables with seaside foraging. They are both square-stemmed, opposite-leaved erect herbs of the mint family and both bear crisp, edible, fingerlike tubers that can be dug and eaten from October until March. I could give descriptions of leaves, flowers, and fruits of these plants, but since you have to recognize them in winter, these descriptions would not be very useful. In winter they will be brown, dry, and shriveled, but you can still identify them by the square stems and the remains of the opposite leaves, and if on digging beneath such a plant you find a cluster of crisp, nutty tubers the size of your little finger, the identification is complete.

There are several plants called Woundwort, but one of these, *Stachys hyssopifolia*, is especially worth your attention; it is very common on Cape Cod and Long Island, in New Jersey, and southward on sandy shores. The sand about the dry, shriveled stems is often full of crisp, elongate, white tubers that are excellent just nibbled raw or sliced thin crosswise and added to a salad. They can also be boiled to furnish an unusual cooked vegetable. The flavor is reminiscent of that of crisp celery, but really it tastes like nothing besides itself, and doesn't need to, for its own individual flavor is good enough for anyone.

The Bugleweeds are *Lycopus* of several species, but the only tubers I have ever been able to try are those of *Lycopus amplectens*, which grows in open sand from Cape Cod to Florida and Mississippi. It grows in other places, too, but in peaty or turfy soils the tubers grow on long stolons and, if found at all, are some distance from the parent plant. In sand, however, the tubers are borne in a cluster right at the base of the dried-up plant and are eminently available to the off-season forager who likes to go food-gathering before the crowds start arriving at the shore. Bugleweed tubers are not as palatable raw as those of Woundwort, but a few sliced into a tossed salad will give a radishlike relish. They are superior as a boiled root-vegetable and also make good pickles if the crisp tubers are merely scraped clean and packed in jars, then covered with vinegar. These two tuber-bearing mints can furnish unusual and acceptable vegetable dishes to grace your wild-food dinners that are given before or after most other wild vegetables are available. I once tried cooking both these tubers together and found the dish delightful. They were washed, scraped clean, then boiled in salted water for 25 minutes. Then they were drained and seasoned with butter and black pepper and served steaming hot. All who tried them wanted a second helping.

SEACOAST ANGELICA AND SCOTCH LOVAGE
The Edible Seaside Umbellifers

The *Umbelliferae*, or parsley family, furnish many of our finest domestic vegetables. Unfortunately, such notoriously poisonous plants as poison hemlock and beaver poison are also found in this family, and since the members of this family are difficult for the amateur to distinguish, I have hesitated to recommend any of the edible wild plants in this group. However, the situation is different for the seaside forager, for as far as I can find out, none of the poisonous umbellifers grows right on the seashore, and two of the perfectly edible species do grow there.

Seacoast Angelica, *Coelopleurum lucidum*, grows on sandy, gravelly, or rocky shores from Labrador to Long Island Sound. It is a coarse plant sometimes reaching a height of 4 feet; at the top it has round-topped umbels 3 to 5 inches across, bearing small white flowers, followed by oblong, squarish, ribbed dry seeds about ¼ inch long. The stem is coarse and green, sometimes dotted with sticky spots. The large leaves are born on green leafstalks that are conspicuously inflated where they join the stem. The leaf is 3-forked, and these forks again subdivided on a plan of three, the final leaflets being egg-shaped in outline, 1 to 3 inches long, coarsely toothed, and green on both sides.

The young stems and the tender stalks of the young leaves are, when peeled, crisp and juicy, with the characteristic odor and flavor of strong celery. This flavor is good, but there is rather too much of it for most American palates, although the raw plant is eaten avidly by natives in several parts of the world. I like a little of it, cut fine, added to a tossed salad, and I like to nibble a peeled stalk as I walk along seashores looking for more important forageables. This celery flavor goes exceedingly well with many kinds of fish and other seafoods, and wherever I have given a recipe that calls for celery, you can substitute Seacoast Angelica by cutting the amount recommended approximately in half.

It is not half bad as a cooked vegetable, even when served alone, and actually, my wife prefers stewed Seacoast Angelica over other wild vegetables I gather along the seashore. Peel the tender stalks and young leaf stems, and cut them in short sections. Cover with boiling water, boil for about 3 minutes, then drain. Add more boiling water and boil until tender. Drain and flavor with butter, salt, and pepper. This plant tastes much like stewed celery, but has a good flavor that is all its own.

Scotch Lovage, *Ligusticum scoticum*, has seldom been used for food on this side of the Atlantic, although around the islands of western Scotland it has long been used both as a raw salad and as a cooked vegetable. In our country, it grows from Long Island to Labrador, although it is rather sparse in the southern part of its range. Look for it along the upper edges of sandy, gravelly or shingly beaches. The leaves and flowers are much like those of Seaside Angelica, but the plant grows in a different manner. Instead of a coarse, central stalk, Scotch lovage has very long-stalked leaves in basal clusters, somewhat like garden celery. These leafstalks are crimson or purple at their broad sheathing bases and are 3-forked, bearing glossy, coarsely-toothed leaflets 1 to 3 inches long. The flowering stems are thin and arching, reaching 1 to 2 feet high and bearing flat-topped umbels of white flowers followed by brown, dry, oblong fruits about ½ inch in length.

The young leafstalks of Scotch Lovage have a clean, aromatic taste, but one that needs to be cultivated. Sometimes one finds plants that have been partially covered by blown-in sea wrack, and naturally blanched, and these are much milder in flavor and more acceptable to most palates. Even those who cannot or will not learn to appreciate raw Scotch Lovage will enjoy it when it is stewed in the same manner as directed for Seacoast Angelica, even though the flavor doesn't resemble celery as closely as that of Seacoast Angelica.

Finally, you can make an aromatic and unusual confection from Scotch Lovage. Cut young leafstalks into short strips, and boil them for about 20 minutes. Drain, then drop into a syrup made of 1 cup sugar and ¼ cup water, and boil for 10 minutes. Drain on wax paper, then roll in granulated sugar and dry for about 2 days, and it will be ready to eat.

28. Sour Sorrel:

SALAD, BEVERAGE, POTHERB, AND DESSERT

(*Rumex acetosa* and *Rumex acetosella*)

THIS useful little plant is also called Sheep Sorrel and Sour Grass in some localities. It has slender halberd-shaped leaves, like arrowheads with wide-spreading barbs, and greenish, unattractive flowers. Sour Sorrel is closely related to the common docks that grow as weeds all over the country and is sometimes called Sour Dock, in addition to its many other folk names.

You will not often find Sheep Sorrel growing right on the beach, but you will seldom have to walk far inland before you find it in great plenty, consequently it is included here. Except in alkaline soils, Sheep Sorrel grows literally everywhere, from the Arctic to our southern border. It is a common weed in fields, along roadsides, and in gardens and flower beds. It thrives in the slightly acid, somewhat sterile soil of some old fields. I never fail to find it in openings and clearings in coniferous forests near the shore.

As seen in the spring and summer, when it is at its best, Sour Sorrel is an inconspicuous plant, usually no more than a thick cluster of halberd-shaped leaves springing directly from the perennial roots. These leaves are small, from 1 to 4 inches long, and one can gather them a handful at a time. If you are not already familiar with this common plant, you will have no difficulty recognizing it by comparing the leaves with the drawing on page 269.

The fact that Sheep Sorrel is edible is not a recent discovery, although it is an almost forgotten one in this country. Back in the days of Elizabeth I, horticultural varieties of sorrel were raised in the vegetable gardens of England, and it was a highly prized food. The French still raise sorrel and use it in some of their superb cooking. In earlier days in this country, housewives sought sorrel to cook as a potherb, while children were familiar with it as a pleasant, acid, nibble and thirst-quencher.

Naturally, over these long years many methods of preparing Sheep Sorrel have been developed. This plant literally can be used throughout the meal, beginning with soup, going on to salad, vegetable, beverage, and, finally, dessert. I wouldn't advise using it this many ways in the same meal, for one can tire of even a very fine food. One of the simplest ways in which I use Sour Sorrel is to add a few of the tender leaves to my sandwiches when I am hiking where it grows. These leaves with their pleasant acidity replace both greens and pickles and are quite good.

To make a hearty camp soup that will furnish a good meal for two, first fry a few pieces of bacon in a deep pan until crisp. Remove the bacon and lightly fry 1 chopped onion and 2 large chopped potatoes in the bacon fat, but do not allow them to get completely done or brown. Now add 2 cups of Sour Sorrel leaves, chopped fine, and the crumbled bacon. If watercress is available, add ½ cup chopped watercress. Cover all liberally with water, and simmer until the potatoes are quite done. Serve in large bowls, and sprinkle the top with finely chopped raw sorrel.

To make a smooth Pureé of Sour Sorrel, cook the leaves in very little water for half an hour, then put through a sieve or strainer. This pureé can be used a number of ways. One of the simplest and best is to put 2 cups of pureé into a saucepan with 2 tablespoons butter and stir well, simmering for 10 minutes. Season with salt and pepper, add 4 tablespoons of cream and stir until it is all smooth and perfectly blended. Serve hot.

When you fillet a nice catch of fish, put the heads, bones, and trimmings into a kettle, cover with cold water, and add 2 or 3 Bayberry leaves, 1 tablespoon chopped wild onion tops, and the top leaves from a small bunch of celery. Cover the kettle, boil very slowly for 2 hours, then strain. This gives you a delicious Fish Stock that blends well with Sour Sorrel Pureé. Heat 2 cups of the stock, add 1 tablespoon of flour blended with a little cold water, and cook and stir until the stock thickens. Add 2 cups of sorrel pureé, ½ teaspoon salt, and ⅛ teaspoon black pepper. Simmer for about 10 minutes, and serve hot with a lump of good herb butter in each bowl. Meat, bone, or vegetable stock, clam broth, rice water, potato water, or barley water can be substituted for fish stock in this recipe and, while each soup will be different, all will be good.

Chopped or shredded sorrel leaves make an excellent garnish and seasoning for fish, oysters, and crab cakes. They are also very fine stirred into potato salad or cooked rice. In salads, sorrel is best used in combination with other vegetables, either wild or domestic, for it is apt to prove too acid for most palates when used alone. A seaside salad I like is made by combining one part sorrel with one part Orach leaves and tips, and a very little Sea Lettuce. This gives three entirely different shades of green and makes an attractive dish. This salad should be dressed with a bland oil and nothing else. The Orach and Sea Lettuce contribute plenty of tasty sea-salt, and the sorrel's flavorsome acidity makes other seasonings unnecessary. Not only is this salad pretty to look at and good to eat, but it is very healthful, furnishing many of the needed vitamins and minerals.

As a cooked vegetable of the spinach type, sorrel can be cooked alone or combined with other wild greens. A mixture we like is equal parts of Sea Blite and Sour Sorrel leaves, boiled about 20 minutes in plenty of water, then drained and chopped. The chopping can be done very handily right in the cooking kettle, if one uses a large pair of kitchen shears. The well-drained, chopped greens are arranged on a dish and garnished with chopped onion, crumbled crisp bacon, and thin slices of hard-boiled egg. Don't salt or season this dish further before you taste it. Sea Blite, gathered along the shore, can be strongly impregnated with sea-salt, and no vinegar is required when you use Sheep Sorrel. This dish can be served either hot or chilled. Another way is to combine 3 cups of the chopped greens with a can of creamed mushroom soup and serve hot as a creamed vegetable.

I found one report that the people of Iceland make a sort of arctic lemonade of Sour Sorrel. The directions merely said that the plant was steeped in water until the juice was extracted. I tried putting some of the chopped leaves in a jar, covering them with water and leaving the whole in the refrigerator overnight. Next day I strained the juice out through a jelly bag. The result was a greenish, sour beverage that wasn't too bad when a little sugar was added. One might be able to learn to like this drink, but don't go telling anyone that I said it was perfectly delicious at the first taste.

The almost fruitlike tartness of Sheep Sorrel suggests that it might be used in desserts in the manner of rhubarb. My one experiment along this line was pretty successful. To 1½ cups of sorrel leaves I added a few dozen mint leaves and ½ cup of homemade maple sugar, then chopped it all very fine. A plain pastry was rolled very thin and cut into 4-inch rounds. On each round of pastry, I placed a rounded tablespoon of the sorrel mixture, dotted it with butter, moistened the edges of the pastry rounds, folded them over, and sealed the edges by pressing around them with the tines of a fork, then perforated a few steam vents. These little turnovers were baked on a cookie sheet in a moderate oven for about 20 minutes, when they were flaky and golden. Served hot, they made a fine finish to a wild-food meal.

29. The Beach Plum

(*Prunus maritima*)

THE Beach Plum, a strange relative of our domestic plums and cherries, actually prefers to grow in semisterile dune sand, and will thrive only in such seemingly poor localities. It is usually found as a shrub, 3 to 6 feet high, though like all seedling fruits it is extremely variable. I have seen Beach Plum bushes 10 feet tall, and others no more than recumbent stems lying across the sand. It is found dotting the dunes from New Brunswick to Virginia, and though usually a seaside plant, it can, where conditions are right—such as in the pine barrens of southern New Jersey—bear tremendous crops of fruit far from the sea. I have picked excellent Beach Plums in the dunes of Indiana, near the southern end of Lake Michigan.

In early spring, the showy white blossoms appear ahead of the leaves and literally cover the branches. At this time the Beach Plum is one of the most beautiful flowering shrubs in all nature. Many a householder coveting this beauty for his own garden has dug up wild Beach Plums and transplanted them into fertile soil under seemingly ideal conditions, only to see them sicken and die, while the wild ones in the poverty-stricken dune sand continue to flourish like weeds.

The leaves that follow the flowers are pointed-oval in shape, from 1 to 2 inches long, with sharp sawtooth edges. The fruit is usually borne abundantly. I have seen such a weight of ripening plums on some specimens that the whole bush was bent to the ground. It takes only a few such bushes to make a great store of some of the finest jams and jellies you ever tasted.

The fruit varies so wonderfully from bush to bush that it is hard to give one description that will cover all that you might find. Typically, Beach Plums when ripe are about ¾ inch in diameter, dull purple in color, with a dense bloom, but I have found bushes bearing pure-red fruits, while others were yellow-orange. In size, you may find Beach Plums a full inch in diameter, while on a nearby bush they may be only half an inch across, and that mostly seed. In quality, the Beach Plum at its best is sweet and pleasant eaten directly from the bush; at its poorest and most inedible in the raw state, it still provides some of the best jam and jelly material to be found growing wild.

To make a clear, tart jelly that is a delight to the eye as well as to the taste buds, crush 5 pounds of fully ripe Beach Plums without peeling or pitting them. Put in a covered kettle, add 1½ cups water, and simmer for 10 minutes. Pour into a muslin jelly bag and squeeze. This should yield about 5½ cups of juice. Add 1 box of commercial pectin to the juice, bring just to a boil, then add 7½ cups of sugar all at once, stirring until it is all dissolved. Bring to a boil again, and boil hard for just 1 minute, then pour into jelly glasses and cover with paraffin, or into half-pint jars and seal with sterilized two-piece dome lids. Never try to double or otherwise increase this recipe. Good homemade jelly must be made in small batches that can be quickly brought to the boiling point.

To make a tart, meaty jam of Beach Plums, the fruit must be pitted but not peeled. An ordinary cherry pitter works well with Beach Plums and makes short work of what would otherwise be a tedious job. To 6 cups of pitted plums, add ½ cup water and simmer for 5 minutes. Then add 1 box commercial pectin, bring to a boil, and stir in 8 cups sugar. Reheat and boil hard for 1 minute, then pour into half-pint jars and seal with sterilized lids.

Beach-Plum jams and jellies are not sickly-sweet spreads to be eaten with toast and tea. They are tart, almost ribald sauces, to be served with the meat course and especially with seafood. Clear Beach-Plum jelly goes with Soft-Shelled Crab as cranberry sauce goes with turkey, or mint jelly with spring lamb. It was a kindly providence that arranged for Soft-Shelled Crabs and Beach Plums to be found in the same localities, and even saw to it that they were in season at the same time of the year.

The usefulness of this excellent wild fruit is not exhausted by jams and jellies. If you complain that most fruit pies are too sweet, you should try Beach-Plum Pie. These little plums can dominate plenty of sugar and come through sharp, tart, and tasty. Even those who lament that they can never make a successful piecrust can turn out a perfect Beach-Plum Pie with my pat-a-pastry. Sift together 2 cups enriched flour, ¼ cup sugar, and 1 teaspoon salt. Add ⅔ cup vegetable shortening. Don't try to substitute cooking oil or butter; use solid, hydrogenated shortening. Cut this into the flour-sugar mixture with a pastry blender or a blending fork, and keep cutting until everything is reduced to the size of fine crumbs or coarse cornmeal. Put two thirds of these crumbs into a metal pie plate and press firmly over the bottom and sides of the plate until you have a smooth crust evenly shaped. Now, to 3 cups of pitted Beach Plums add 2 cups sugar, ¼ cup enriched flour, and ½ teaspoon almond flavoring. Mix well, pour into your patt.d crust, dot the top with 2 tablespoons butter, then sprinkle the rest of the flour-shortening crumbs over the top to serve as a top crust. Bake 40 minutes in a preheated oven at 425°. This pie is especially delicious if taken from the oven just in time to be served still faintly warm.

My own favorite desserts are the fluffy, light, chiffon-type pies, and I have learned to make chiffon fillings from wild strawberries, wild cranberries, persimmons, pawpaws, and even from a mixture of wild Jerusalem artichokes and chopped hickory nuts. However, some of my guests complain that my chiffon pies are too sweet and my Beach-Plum Pies too tart. This suggested bringing these two together to produce a pie that would be absolutely perfect. I washed about a quart of whole Beach Plums, added 1 cup of water, simmered them until they were soft and tender, then pressed the pulp through a sieve to remove the stones and skins. This sieved Beach-Plum pulp became the base for one of the finest pies I ever concocted.

To make this ultimate pie, first mix thoroughly ½ cup sugar, 1 envelope unflavored gelatin, and ½ teaspoon salt. Separate 4 eggs, save the whites, and beat the yolks with 1 cup Beach-Plum pulp and ¼ cup water. Mix this with dry ingredients, then cook and stir over medium heat until the mixture comes to a boil. Remove from heat and cool until it sets enough to mound slightly when spooned. This

Beach Plum (*Prunus*).
Left, *in blossom:*
Right, *in fruit.*

Bayberry (*Myrica*).

takes 30 minutes or more, giving you time to make the corn crust which is the perfect container for this perfect pie. To 1 cup of finely-crushed corn flakes add ¼ cup sugar and ⅓ cup melted butter, then stir and mix until everything is evenly dampened. Press this firmly onto the bottom and sides of a 9-inch pie plate and shape into a smooth crust. Bake only 6 minutes in a preheated 375° oven, then cool.

While the crust is cooling, beat the egg whites until they form soft peaks, then gradually add ¼ cup sugar while continuing to beat until there are stiff peaks. By this time the plum-yolk mixture should be just right. Be sure it is stiff enough to mound a bit when spooned, but don't let it get too thick or you'll have a lumpy pie. Test it often and catch it when it is just right, then fold into the egg whites, using gentle but thorough strokes. When all is smooth and evenly colored, pour into the corn crust. The filling is generous, but stiff enough so that you can pile it into a mountain of a pie with decorative swirls and peaks at the center. Chill until firm, then serve in mile-high wedges. The Beach Plums impart an inviting pink color and a piquant sprightliness to the smooth, fluffy chiffon filling that seems to be universally appreciated. When I first tasted Beach-Plum Chiffon Pie I decided to cut from my list any acquaintance who disliked it, and I have yet to lose a single one on that basis.

If Beach Plums are plentiful, there is no reason you cannot enjoy this special pie all year. Make the pulp as described above, pour it into half-pint jars, seal with sterilized dome lids, and process in boiling water for 20 minutes. These will keep until Beach Plums are ripe again, and each jar will make a memorable pie.

30. The Ubiquitous Bayberry
(*Myrica* species)

THE Bayberry, also known as the Wax Myrtle, Candleberry, Tallow Bush, and a host of other regional folk names, is common on sandy coasts from Nova Scotia to Florida, and is even found around Lake Erie. It is a real lover of the sea, and often grows where the highest spring tides lap its base. Asa Gray, in his original *Manual of Botany*, was content to lump all the kinds of Bayberries under the one specific name, *cerifera*. Later workers in the field have divided them into a half-dozen or more species, the number largely depending on the botanist doing the classifying. The differences between these closely-related species are largely technical and mainly of interest to taxonomists, and all can be used in the recipes and processes given below.

Although usually encountered as a shrub 3 to 8 feet tall, the Bayberry has been known to break out of these size limits and surprise everyone by growing into a tree as much as 35 feet high. It has stiff branches covered with gray bark. The leaves are narrowly oval, usually with entire margins, but often toothed toward the outer end. The leaves are in erect clusters on the outer ends of the branches. Just inward from the leaf clusters are found the fruiting stems bearing clusters of grayish-white bayberries in autumn and winter. These so-called "berries" are really dry drupes or nutlets about ⅛ inch in diameter, and they derive their color from a thick coating of granular wax that is pleasantly aromatic. The leaves also have an attractive fragrance when crushed.

Today, Bayberries are best known for the fragrant, green candles made of its wax and commonly sold in quaint little gift shops all over Cape Cod. To our ancestors, the Bayberry bush was not only a source of candles, when candles were utilitarian lights rather than mere dinner-table decorations, but it was also a medicinal herb, a dye plant, and a kitchen herb or spice; the wax was used as an aid to ironing clothes and as an ingredient of homemade soap, as well as for making candles. Since this is a book about foods, we are chiefly interested in the culinary uses of Bayberry, but it will do no harm to learn a few of its other virtues as well.

It is easy to prepare a supply of Bayberry leaves for use in cooking. Just gather a few clusters of the leaves any time during the summer or fall when you're visiting the shore. Spread them on trays and dry them indoors, at room temperature, then seal them into a jar with a tight-fitting lid to keep the flavor in, and set it on your herb shelf. I will not give recipes for its use here, as throughout this book you will find recipes calling for Bayberry leaves as part of the seasoning. Generally, any crab-boil, stock, broth, soup, stew, or chowder is improved if one or two Bayberry leaves are boiled with it and removed before serving. In the literature on this subject, one finds the Bayberry leaf referred to as a "substitute" for the tropical bay, but this it is not. The flavor of Bayberry is quite different from that of commercial bay leaf, and in my opinion much better. A Bayberry leaf can impart its delicate aroma and subdued flavor without dominating the dish, and I consider it one of the finest herbs on my pantry shelf.

Why not make your own Bayberry candles? They are lovely as gifts, exuding the very odor of Christmas. Then, what could be more appropriate at a "wild party" where you have foraged the food from seashore, forest, and field, than to light the dinner table with Bayberry candles made with your own hands? The waxy berries are not nearly as tedious to gather as you might think, although it does take an awful lot of berries to make a very few candles. Boil the berries in a deep kettle for about 10 minutes, then strain the boiling water with the melted wax through a cloth strainer into a straight-sided crock. This removes all trash and the expended berries. As the liquid cools, the wax hardens in a cake on top of the water and can be broken, removed, and remelted for making candles.

Some like to make their candles in fancy molds, but I much prefer the old-fashioned, hand-dipped taper. A word about wicks: a candle wick is much more than just a piece of string. Candle-wicking is cleverly braided with one tight strand, so that it will bend over and more completely oxidize as it burns, and it has been pickled in a mineral solution so it will char and drop off, rather than smolder and smoke. To be sure of candles that will burn beautifully without constant attention, you had better buy your wicks from a craft supply house and not attempt to manufacture your own. Melt the wax in a deep container, such as a tall fruit-juice can. By tying your wicks to a 4-pronged twig you can dip 4 candles at once. Keep the wax just above the melting point and lower the wicks into it, then suspend them directly above the can and allow the wax to set. As soon as they are well hardened, dip them again, and continue until the candles are the desired size. They won't build up perfectly round and smooth, but the unevenness of hand-dipped candles only enhances their beauty. When they get too much out of shape the higher ridges can be shaved off with a sharp knife just before the final dip. Shave the bases so they will fit your candlesticks, and your candles are finished. They will be a grayish blue-green in color and I prefer them natural, but you can change them by using food colors in the melted wax, if you want to get a more calculated effect.

The perfect candlestick for these wild candles is made of a short, thick section of white-birch branch. Smooth one long bark side so the branch rests steadily, lengthwise, on the table, and drill a hole—or holes—of the proper size on the upper side. Staple a couple of sprigs of holly or other evergreen on either side of the hole, or holes, insert your candle, and you will have a beautiful, woodsy light that will excite only favorable comment.

When soapmaking was a universally practiced household art, the housewives along the Atlantic shore had in the Bayberry a never-failing source of "soap grease," for Bayberry wax will saponify like fat. If you would like to try your hand at manufacturing Bayberry soap, a sample batch of a few bars can be made of a single cupful of wax left over from your candlemaking. Just dissolve 2 tablespoons of sodium hydroxide, commonly called lye, or caustic soda, in ⅓ cup of cold water. Use only glass, iron, enamelware, or stainless steel containers while making soap; never use aluminum. The lye will cause the water to heat up. Let it cool to about lukewarm, but be sure you feel only the outside of the container, or you might end up wondering what became of your finger. Pour the warm lye solution into 1 cup of barely melted Bayberry wax and stir for 20 minutes. Then pour into molds, and keep in a warm place for 3 days. The fatty acids of the Bayberry wax completely neutralize the caustic principle of the lye during saponification, and you are left with a mild, free-lathering white soap that has a faint fragrance of Bayberry.

Another domestic trick of our great-grandmothers was to rub a piece of Bayberry wax over their warm flatirons. This made the irons glide smoothly, and imparted a subtle hint of Bayberry perfume to the ironed clothes.

Nor did our frugal ancestors throw out the water in which they boiled Bayberries to extract the wax. Woolens gently boiled in this water were dyed a dull, but pleasing, blue, and this color was fast and washable without further treatment. The water was also a home remedy for several illnesses, being used as a gargle in cases of sore throat; taken internally, an ounce or 2 at a dose, for diarrhea; and used as a mouthwash to toughen the gums. Now that we know the many ways in which the Bayberry contributed to the comfort of our forefathers, and the many ways it can add to our own enjoyment, I'm sure we'll find it harder to overlook this humble shrub of sand flat and shore.

31. Catching and Cooking the Wild *Et Cetera*

YEARS ago, when this book was just a nebulous idea, I wondered if there was sufficient material in this field to fill a decent-sized volume. This proved to be no problem. Each time I started exploring the edibility of one seaside creature or plant, I stumbled across the trails of at least two more with possibilities, and so I end this book with more unwritten material than I had when I started. The problem became one of deciding which parts of the vast store of interesting materials to exclude. No man who has access to the sea and shore should ever die of starvation. I ran out of space long before I ran out of seaside foods to talk about, but before I leave this subject there are some others I simply must mention. For instance, not nearly enough has been said about Lobsters.

The common Northern Lobster, *Homarus americanus,* is easily the most important crustacean inhabiting the coastal waters of the North Atlantic States and the Maritime Provinces of Canada. In this area lobstering is a major industry, with millions of pounds being caught and sold annually. Stand on almost any headland in Maine and try to count the Lobster buoys that can be seen through a pair of binoculars. The chief problem that confronts a Maine Lobster is deciding which of thousands of Lobster pots he will walk into when he reaches marketable size. The end product of this industry is universally recognized as one of the finest of all seafoods, the very name *lobster* being synonymous with luxurious dining, except to those unfortunate few who are allergic to its flesh.

I was surprised to find that lobstering, besides being an important industry, was also an outdoor sport or recreation of no mean order. I was formerly under the impression that lobstering was the monopoly of salty old professionals whose wisdom in the ways of boats and the sea was handed down through generations, and that breed actually does exist, God bless them. However, there are others, just as salty and seagoing in appearance, who engage in this activity as a part-time occupation or hobby. Among them, the love of boats, the challenge of weather and sea, the fishing thrills, and the satisfaction of furnishing their own tables with an abundance of this delicacy, all are more important than the money they make from selling their surplus. Not that the money isn't welcome; it pays for boat and equipment and furnishes luxuries they could not otherwise afford, for unlike most hobbies, lobstering can be made to pay. Many of these semiprofessionals work at other jobs and do their lobstering before or after work, and on weekends. Some are summer people, who make lobstering pay for a long vacation by the sea, and others are retired men looking for an outlet for their surplus energy.

A few of these part-timers have discovered that Lobsters can still be caught in the old-fashioned way by using hand-pulled traps and ordinary rowboats, and substituting hard work for fancy equipment. On one cove I visited in Maine, there lived a partly disabled veteran who eked out an inadequate pension by lobstering. He used a sprit-rigged dory that he sailed when the wind was right and rowed when it wasn't. Daily he pulled dozens of traps with no power but his own strong arms and shoulders. He knew the cove as well as he knew his own front yard, and after finishing for the day he would sometimes row to a favorite fishing spot and spend the afternoon catching a supply of mackerel or flounder to sell, eat, or store in his freezer. Despite a limp, he cultivated a small vegetable garden behind his cabin, and with the aid of his freezer he had all the fresh vegetables he could use the year around. When the wild blueberries, raspberries, and shadberries, that grow so plentifully in Maine, were ripe, he would gather as many as he could use and preserve or freeze a supply for off-season use. He had very little money to spend, but the wealthiest gourmet couldn't have bought better food than he habitually ate, and he is one of the few persons I have ever met who was living his life exactly as he wanted to.

Then there are the wonderful boys of that coast. I used to fish from a cannery wharf right next to a Lobster pound, and besides the professionals, with their large expensive boats, I would see a goodly number of teen-age boys, with outboards, rowboats, or beat-up motorboats, pull in and unload sizable catches of Lobster. One of these boys would sometimes join me on the wharf. He was a skilled handline fisherman, and from the flounders and pollack he caught he would cut nice boneless fillets to take home, throwing the rest of the fish into a basket to use as bait for his Lobster pots. He told me that lobstering not only financed his boat and its motor, but paid his school expenses, bought his clothes, and supported a hot-rod as well. He also contributed to the family table all the Lobster, fish, crabs, and clams they could use.

He worked hard but was far from being a joyless drudge, and he certainly wasn't a "square." With the extra money he made, he could dress as cool as any city cat, and did, hairdo and all, and he always had some money to jingle in the pockets of his skin-tight pants. He kept his boat immaculate, and had bright-red marine upholstering on the forward seat. He was very popular with the girls, who vied for an invitation to go out with him in the boat. Nearly every day as he pulled in to unload his catch he had a new girl with him, each seeming prettier than the last. Professional lobstermen pull their traps with a motor-driven winch, but when I asked him if his boat was equipped with a winch, he looked at me slyly and said, "No, but it's usually equipped with a wench."

One day when he had no feminine companion he invited me to go along. That day I saw that this boy not only had a good thing, financially and socially, but that he was enjoying life more than most. There was an excited anticipation at the pulling of each trap, a cry of joy if it yielded one or more marketable Lobsters and a groan if it was empty, and his zest for the game was as keen at the last trap as at the first. He took pride in placing his Lobster pots in almost inaccessible nooks and coves where the deeper-draft boats were afraid to go. His life was hazardous enough to furnish the thrills that boys seem to require, but his were meaningful hazards, not the senseless thrills of drag racing and "chicken" driving. He was a superb boatman and knew every mood and change of the sea. He kept his own boat and motor in good repair, and had learned many of the skills of the marine mechanic and boatbuilder. The outdoor life had given him splendid health and a fine appetite, and hand-pulling dozens of Lobster traps each day had developed magnificent arms and shoulders. He had more money and more fun than do most high-school boys, and he was learning a self-reliance and independence that is rare in modern youngsters. He was no sissy-pants or goody-goody type, but he was a fine boy in the process of becoming a fine man. Lobstering as a hobby or part-time occupation has much to recommend it besides the obvious value of yielding an abundance of one of the finest foods that ever came from the sea.

Although most people think of the Lobster as a creature of the North, it is actually found from Labrador to Virginia, and there are good lobstering places as far south as New Jersey and even Delaware. I have a good friend in southern New Jersey who owns a boat and loves fishing, and I prodded him into adding lobstering to his sports. He agreed, on the condition that I was to build the Lobster traps, which I did. The first week these traps were in the water they didn't catch a single Lobster, but my friend was highly pleased nevertheless, for each day he found from one to four nice sea bass in nearly every trap. He has since found where the Lobsters lurk, and although the traps still function better as bass traps than as Lobster traps, I am occasionally recompensed for the labor of building them with some delicious Lobsters fresh from the sea.

Lobstering is the perfect hobby, but unfortunately it is a hobby that can only be ridden by permanent residents of the states that border on the North Atlantic. Strictly enforced laws prevent non-residents from catching Lobsters under any circumstances. As I live in Pennsylvania, I can only practice this hobby by proxy, watching others thrill to a sport from which I am excluded. I am not suggesting that lobstering should be unregulated and open to all comers—that would be disastrous—but I am hinting that some Lobster state could make a very good thing of opening limited, licensed, and strictly regulated lobstering to nonresident sportsmen.

While I was vacationing in Maine, nearly every tourist with whom I talked was fairly aching to catch a Lobster of his very own. Lobstering is Maine's third largest industry, and the state very understandably wants to preserve and protect this source of income for its

people. However, tourism is Maine's second largest industry, and if
the state would open a season when licensed sportsmen could catch
some Lobsters for their own use, this industry would automatically
expand.

I doubt that a single professional lobsterman would suffer from
such a provision. They would still have to furnish the Lobsters for
the market, for, as is now the practice with other game laws, the
sportsman would be forbidden to sell his catch. During the sports
season, some lobstermen might prefer to rent out Lobster pots, char-
ter boats, and act as skippers and guides for the sportsmen, and they
would undoubtedly find this a profitable sideline. To prevent a de-
pletion of the Lobster supply, there could be strict bag limits and
reasonable license fees for the sportsmen. The license fees could go
into a conservation fund to combat pollution and increase the areas
where Lobsters abound.

Let Maine calculate how much revenue its state and people receive from each Atlantic salmon and striped bass caught by nonresident sports fishermen, or how much money comes into the state
for each deer or bear killed by nonresident hunters, and they will
know that they only have to open a sports lobstering season to reap a
rich harvest of tourist dollars. All that is necessary is a slight shift
in the way the state conservationists think about Lobsters. Instead of
thinking of the Lobster solely as a marketable crustacean, and lobstering only as an industry, let them recognize that it would be very
profitable to think of at least some of the Lobsters as game and amateur lobstering as a sport.

The Spiny Lobster of the South, *Panulirus argus*, is not closely
related to the Northern Lobster, but is a wonderful creature on the
table, nevertheless. It resembles its northern counterpart except that
the Spiny Lobster lacks pinching claws and has a carapace covered
with nodules, warts, and sharp spines. The size is 8 to 16 inches
from the nose to the end of the telson, and the coloring can only
be described as variegated, with patterns ranging through the reds,
browns, yellows, and blues. It is essentially a tropical creature, being
found around the world along the shores of warm seas. On our own
shores it is found from North Carolina to Florida, with an almost
identical species, *Panulirus interruptus*, on the West Coast, which
ranges from Santa Barbara into Lower California. Besides these loca-

tions on mainland shores, the Spiny Lobster is abundant around those favorite playgrounds of American vacationers, Bermuda, Puerto Rico, the Virgin Islands, and Hawaii.

The game laws concerning the Spiny Lobster are not so strict and exclusive as those that prevent outsiders from taking a Northern Lobster, so I have had firsthand experience with this creature. In Hawaii, I was, for a time, a professional fisherman of Spiny Lobsters. We caught them for market in 6-inch-mesh nets that were only 4 meshes high and sometimes a thousand feet long. These long nets were laid along the bottom among the coral heads and left down overnight. In running the nets the next day, it was frequently necessary to dive down and free them from coral branches, but in that warm water, it was no hardship.

We also caught a few Spiny Lobsters in traps. We set the traps for fish, but Lobsters persisted in walking into them. As a sort of postman's holiday we sometimes went fishing for Lobsters at night, purely for sport. "Fishing" is the right word, for we actually caught Lobsters with a hook and line. The Spiny Lobster comes out at night to feed, and we caught them from rocky ledges and tiny islets, using a line baited with a skinned octopus tentacle, or the foot muscle of almost any shellfish, on a tiny 3-pronged hook. The lines were attached to long bamboo poles. The Lobster actually takes the bait in its mouth, and is hooked like any fish, and then must be brought ashore in one long sweeping motion. It was eerie fun landing these scuttling monsters in the dark. Most of those we caught weighed from 1 to 2 pounds, but one night a Hawaiian girl who was fishing with me landed a 9-pound giant. Sometimes we found a fierce moray eel on the hook instead of a Lobster, but we didn't complain, for I had learned to enjoy a plate of hot, broiled eel after an arduous night on the reefs.

All these ways of catching Spiny Lobsters are fun, but the way to wring the maximum amount of sport from this creature is to dive down to the bottom and catch it with nothing but your hands. Skin diving is growing in popularity every year, and anyone who has ever dived around a coral reef knows why. Just to go down to the magnificent underwater world with its gardens of colorful plants with myriads of gaudy little fish swimming among them, is reward enough, and the Lobsters come as a pure bonus. With scuba-diving gear the

Right, Shrimp.
Below, American Lobster.

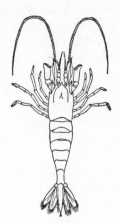

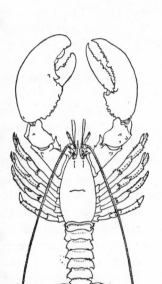

Right, Spiny Lobster.

Spiny Lobsters are practically at your mercy, but I have taken many a fine Lobster with no more diving equipment than an ordinary snorkel mask. Diving should be done in the middle of a sunny day, when the light below is best.

Wear a pair of heavy gloves to protect your hands from this Lobster's sharp spines. You may carry a spear, but use it only on those that are in holes too deep for you to reach them. When you get a Spiny Lobster in your hands, hold on for dear life. It will flap that strong tail mightily, and set up a terrible commotion, and the first time this happens most neophyte divers panic and let go. It is all bluff, and you can soon have your Spiny Lobster safely into a bag if you just hold on.

Be careful how you stick your hands into holes in a coral reef, for pugnacious tropical fish and eels lurk in such holes and could easily amputate one of your fingers. The Hawaiians say that if a Spiny Lobster seen in a hole has one antenna forward and the other backward, it is a sure sign that there is a dangerous eel in the hole behind it. Tie some strips of cloth to the shaft of your spear and shake these in any hole you are afraid to reach into. This will frighten the Lobster and cause it to come dashing out into reach.

Although I lack experience with mainland Spiny Lobsters I feel certain that all these methods will work along our own southern shores. The Spiny Lobster can also be caught in ordinary lobster pots such as are used to take the Northern Lobster. Recently, on a rocky coast in Southern California, I found a magnificent Spiny Lobster stranded in a tide pool, and merely waded in and picked up the makings for a wonderful Lobster dinner. It was like receiving manna from heaven. The Spiny Lobster can be cooked in any way you would prepare the Northern Lobster, and you will find it delicious.

Any book about seafoods should have more to say about the Shrimp than this one has, so far. It is not that I am prejudiced against Shrimps, indeed I have a passion for them when they are skillfully cooked and properly served. The trouble is that I have never figured a way to catch my own supply of Shrimps, and this book is not about the many fine seafoods that can be bought in the markets. I know the Cajuns of Louisiana and Mississippi catch Shrimp for their own use, but I have never had an opportunity to study their methods. Some day ... but that will be in another book.

Another crustacean that deserves more mention is the Mantis Shrimp or Squilla. Despite the fact that one of its common names is Mantis Shrimp, it is not closely related to the Shrimps, but belongs to an entirely different order and family. There are more than 200 species of Squilla, but the whole order is a closely unified group, and the species look much alike to the layman's eye. The *Squilla empusa*, common from Cape Cod to Florida, becomes 8 to 10 inches long, and slightly resembles a slender, misproportioned lobster, but has many features that are its very own. There are no chelipeds, or pinching claws, but it does carry a pair of fierce-looking toothed arms that it holds in a prayerful attitude until they are needed to grasp and kill some unsuspecting prey. These praying arms resemble those of the praying mantis, hence the common name, Mantis Shrimp. It has a short thorax, or forward part of the shell, and a long abdomen ending in a decorative and complicated caudal appendage, or telson, which is the widest part of the whole slender creature. The body is light green, bordered on the after-edge of each segment with dark green with a narrow edging of bright yellow. The telson is tinged with rose, yellow-green, and black.

It was in Hawaii that I first met the Squilla, not the same species as our local ones, but very similar, and there I learned to appreciate its delicate and fine-flavored flesh. A small Hawaiian boy, bribed with candy bars, taught me how to snare the Squilla. For the privilege of sharing my dinner, he also showed me how to clean and cook them.

These Hawaiian Squilla were about 6 inches long. When I returned to the mainland I was delighted to find even larger and finer Squilla along our own East Coast, and was even more delighted when I discovered that our local Squilla have no objection to being caught in a Hawaiian snare.

The Squilla live in shallow burrows in stiff mud, at, or just below low-tide level, but it is useless to try to dig them out, as the burrows have lateral passages and several exits. To catch them, fasten a piece of bait securely to the end of a wire or slender switch. The tough foot muscles of mud snails or Marsh Periwinkles, or a small bit of clam or squid, make excellent baits. Make a sliding loop in the end of a piece of fine wire and fasten this loop to the end of a stick about 2 feet long. Lead the running end of the wire up the side of the stick and make a loop for your finger near the top. Place the lower loop around the mouth of a mud burrow likely to harbor a victim, and with your other hand thrust the baited wire or switch down the hole. If the Squilla is at home you will immediately feel an almost electric vibration as it attacks the bait furiously. Slowly withdraw the bait, and the Squilla will follow it up right into your loop. Pull the loop tight with your finger, and you have mastered the fine art of lassoing Squilla. It takes patience to catch enough for a meal, but catching your own is the only way you are ever going to be able to taste this epicurean seafood, as Squilla does not appear on any market that I know of.

It seldom pays to seek local information about the Squilla population of any area. So inconspicuous do these creatures remain in their lowly mud burrows that, even where they are abundant, local people may be unaware of their existence. Even those who know there are Squilla about may not know that they are edible, for few people have learned to catch them in sufficient quantities to make cooking them worthwhile. The ingenious and wide-awake beachcomber can always find a supply of Squilla, if any at all are about, by poking a baited wire down any likely-looking holes he can find in the mud at low tide.

To clean Squilla, dump them alive into boiling water, which kills them instantly. Let them cook about 2 minutes, or until the shells turn bright red, then run cold water over them until they are cool enough to handle. Break off and discard the thorax, as it contains no edible meat. With a pair of sharp-pointed shears, snip the shell that covers the abdomen along the back from end to end. The shell will then come away in one piece, leaving a finger-sized piece of meat. Your shears will have made a groove down the center of the back, and in the bottom of this groove you will see a dark vein that should be removed. This is very important, for if this mid-vein is not removed, the Squilla will taste muddy. Wash the meat clean under runnng water and prepare it in any way that you would use fresh shrimp. If they are to be used in salads, cocktails, or eaten as plain boiled Squilla (there are worse dishes), then allow them to boil 10 minutes before draining, cooling, and shelling. A Mantis Shrimp cocktail is a fine way to begin an unusual seafood meal. Just dice the boiled Squilla tails and cover with seafood cocktail sauce.

Short-cook the ones that are to be fried, scalloped, broiled, or sauced.

Fried Mantis Shrimp is like fried shrimp, only more tender, delicate, and delicious. Dip the cleaned tails in the frying batter described on page 232, and fry to a light golden brown in bland oil heated to 375°. Squilla also make a marvelous Newburg, if you follow the recipe on page 99, and they fully deserve the corn-cream pancakes and flaming brandy described in that same section. Cook Squilla in any of these ways, and I'm sure that from the first taste you will agree that they are worth all the effort it takes to catch and prepare them.

It would be impossible to describe all of the hundreds of species in the phylum Mollusca that can be used for human food, but a few more must at least be mentioned. Turban Shells are abundant along the West Coast, and are too delicious to be ignored. As the illustration shows, these are snail-like gastropods of characteristic turban shape. The Black Turban is from 1 to 1½ inches in diameter, ranging from southern Alaska to Mexico, and the Brown Turban, about the same size, is found from northern California to Mexico, thus, in the southern part of their ranges these two species overlap. Look for them on rocky, semiprotected shores at low tide. At night, and on cloudy or foggy days, they tend to congregate in crevices and on the sides of boulders. You can often gather enough for a real gourmet meal from one such cluster.

Don't be surprised if you see some of these shells walking across the bottom of a tide pool. If you examine one of these walking shells you will find that the original inhabitant has long since gone to its reward, and the shell has been appropriated by a hermit crab. It is still a debated question whether the hermit crab waits about decently for the Turban to die, or whether it has something to do with this death, thus gaining an enormous meal and a new home in one operation. These hermit crabs are, in turn, sometimes invaded by parasitic barnacles that live on them and may eventually cause their death.

While gathering Turban Shells, you may notice that many of the shells have tiny, limpetlike brown shells clinging tightly to their outsides. This is the little slipper shell, *Crepidula adunca*, a hitchhiker that is merely along for the ride. Their presence does not harm the host, nor make its meat unfit for food. Scientists have found other, even tinier, parasites on the parasitic slipper shells, and other, still smaller parasites on these parasites. This parasitic habit of sea creatures often reminds me of the verse that goes:

"Great fleas have little fleas
Upon their backs to bite 'em,
And little fleas have lesser fleas,
And so *ad infinitum*."

I learned from some Italian émigrés in California to gather the Black and Brown Turban Shells and boil them in hot oil for about 15 minutes. This in-the-shell French frying bubbles some of the water from the flesh and shrinks the creature enough so that the meat is easily slipped from the shell. They are served in the shell, and each diner picks out the meat with a bent pin or a nutpick. This sounds somewhat greasy but is, in fact, not offensively so, and these boiled-in-oil shellfish are delicious with unbuttered, crusty Italian bread or plain, fluffy rice.

Another way is to boil the creatures, shells and all, in very salty water for 20 minutes. The salt is not for seasoning, but is needed to shrink the tissues and make the meat easy to remove from the shells. If you or your guests object to eating "snails," remove the meat from the shells, and prepare it according to any of the recipes recommended for Periwinkles.

The very similar Red Turban, *Astraea gibberosa*, 1½ to 3 inches in diameter, found from Vancouver Island to California, and the slightly larger Wavy Turban, *Astraea undosa*, from northern California to well down into Mexican territory, will be seen while you are gathering the Red and Black Turbans. These *Astraea* species are called Top Shells in some parts of their range, because the shape resembles the wooden tops children spin. You can identify these Top Shells from the drawings on page 307. The Italians who taught me to eat Turban Shells did not eat these Tops, and I didn't know they were edible until I saw them canned in a Chinese grocery store. The canned product was delicious, with a flavor somewhere between those of oyster and clam, with a few gustatory highlights that were its very own.

After tasting the canned product, I took all the fresh Top Shells I could find. They are larger and need a bit more cooking than the

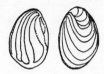

Wavy Turban Shell (*below*).
Size: *about 1½ to 3¾ inches in diameter.*
Above, operculum (*trapdoor opening*).

Above, Brown Turban Shell.
Size: *about 1 to 1½ inches in diameter.*

Black Turban Shell. *Size: about 1 to 1½ inches in diameter.*

Chestnut Turban Shell. *Size: to about 1½ inches in diameter.*

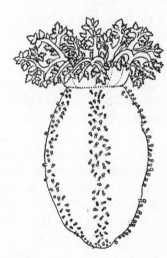

Sea Cucumber

Turbans. Boil them in very salty water for 30 minutes, then remove the meat from the shells. It is very good just diced and added to salads or seafood cocktails, or it can be cooked with eggs, scalloped, or prepared by almost any of the Whelk or Periwinkle recipes, and you will find it better than Whelk. For a dish with an Oriental touch, lightly brown diced Top Shell meat in peanut oil, then add 1 tablespoon soy sauce and 1 teaspoon grated, fresh, green ginger. Serve on steaming plates of fluffy rice.

The Turban and Top Shells of the Atlantic shores are mostly too small, or live in too-deep water to interest the food-seeking beachcomber, but there is one eastern Turban I would like to try. This is the Chestnut Turban, *Turbo castanea*, which reaches 1½ inches in diameter, and is found from Carolina to Florida. I have not had an opportunity to test it on the table, nor have I heard of anyone eating it, but I would appreciate hearing from any good cook in the South who has experimented with this promising seafood.

A book about edible seashore life would be remiss if it didn't say more about that very peak of molluscan evolution, the Octopus. I realize that most Americans are filled with horror at the very thought of eating Octopus, but this is no more than a sectional prejudice, for the French, the Italians, the Polynesians, and the Japanese all consider Octopus very fine fare indeed. Even those Americans who have given this food a fair trial invariably find it delicious. Few laymen realize that the mobile and active Octopus really belongs to the phylum Mollusca, thus making it a relation of the sedentary clams and oysters. Its flesh is just as edible as are those delicacies.

I first encountered Octopus as a food in a native snack bar in Honolulu. I had gone in with a Hawaiian friend, who ordered a dish of *puloa* as a snack. This turned out to be a plate of tender little disks of meat that we speared with toothpicks. It was indescribably delicious. It was only after I had finished the second dish that my companion informed me that I had been eating Octopus, but by that time I was a firm convert and have never refused Octopus since.

The Octopus is conceded to be the most highly developed and most intelligent of all invertebrates. Although it is a mollusk, the shell is absent or vestigial, and with no limiting skeleton it can squeeze through openings that appear far too small to allow it to pass. It has eyes that are as highly developed as our own, and it has the largest and best-functioning brain of any invertebrate animal. The 8 prehensile arms give it an agility and dexterity rivaled by no other creatures except perhaps men and monkeys. The Octopus has been observed picking up small stones and slipping them into the open valves of unalert oysters and clams, wedging the shells apart so they could not be closed, so the Octopus can feast on fresh mollusk meat at its leisure. This use of tools shows not only great dexterity but a high order of intelligence.

The Octopus has a chameleon-like ability to change its color and can blend so perfectly into any background that it becomes practically invisible. I have seen an Octopus turn deep pink after being speared, but whether it was flushing with anger or blushing in embarrassment at being outwitted, I cannot say. The prehensile arms also double as legs and the Octopus can move at a good rate over rocks or sand on the beach. Once in the water, the arms straighten out opposite the direction of travel in an efficient streamlining, and the Octopus propels itself at great speed with a jet of water ejected from its mantle. On the under-side of the arms are two rows of suction disks so powerful that even a small specimen is very hard to pull from a rock or the side of a boat, once it has gotten a firm hold. When threatened, the Octopus can eject a dense, sepia-colored fluid into the surrounding water and escape under cover of this "smokescreen." Jet propulsion, streamlining, intelligence, good vision, agility, dexterity, use of tools, ability to change color, use of smokescreens, these are all unexpected attributes in a mollusk, yet here they are, all in the same amazing creature.

The folk name most often heard for the Octopus is Devilfish, a derogatory common name it shares with the giant ray of the tropics. I sometimes wonder if the horror with which we commonly view the Octopus is not a little mixed with unadmitted fear that this creature might someday supplant us. The rounded "head" and large eyes give the Octopus just enough resemblance to a deformed human being to make it extremely repulsive to our sight. Seen not as a noseless subhuman creature, but as a highly evolved mollusk, the Octopus is beautiful.

To justify our unreasonable fear and horror of this creature, we have built a whole literature of falsehood about its really nonexistent ferocity. There are literally hundred of stories about Octopuses attacking divers, holding swimmers until the tide came in and drowning them, and even overturning boats to get at the occupants. I would hesitate to say that every one of these stories is pure fabrication, but I have caught dozens of Octopuses, and have seen hundreds caught, and I have yet to see a diver or fisherman even slightly endangered by this creature. The Octopus is clever, evasive, agile, and strong, yes, but dangerous, no. It might be humiliating to my own species, but it is my opinion that the Octopus views human beings with the same kind of horror that most human beings view it—and with good reason. A creature that has been accustomed to dine on the very finest of crustacean and molluscan shellfish would hardly find ordinary human flesh palatable.

In Hawaii, we caught Octopus by dragging shiny lures made of cowrie shells across the bottom. We also speared them on the reef during the day, or went out at night and speared them by torchlight. We sometimes speared Octopus while diving for Spiny Lobster or fish. In California, I have caught them with spears and gaff hooks by poking around in likely holes at extremely low tide.

The small Octopus, *octopus bimaculatus*, of Southern California and the Pacific shores of Lower California, is the species most available to the Western beachcomber, as it comes into the intertidal zone. The easiest way to catch this Octopus is to turn over stones on the outer tide flats during unusually low tides. If an Octopus is found under a stone it must be grabbed quickly with bare hands and put into a wet burlap bag. One low tide should yield all the Octopus you can use, and if it is an exceptionally low tide, you can even find good Octopuses near the metropolitan areas of Los Angeles and San Diego, while down in Lower California they become abundant on rocky shores. They are around all year, but are most plentiful in April and May and again in November and December.

Octopus rugosus is the common Octopus of the Florida reefs and is sometimes found as far north as the Carolinas. I have not taken this Octopus, but my nephew has often speared them while skin diving and I'm sure they would respond to the same methods we used in Hawaii.

To clean an Octopus turn it upside down, sever the few muscles that hold the viscera, then turn the "head" inside out. Remove the viscera and the dark ink sacs that you will find inside. Turn it right

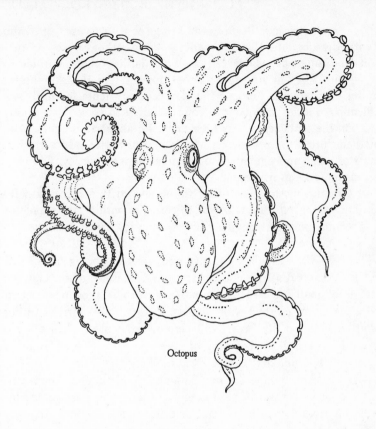

Octopus

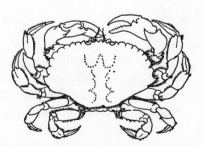

Rock Crab (*C. irroratus*). Size: about 3 by 4 inches.

side out again, place it in a pan, and cover it with coarse salt. Scrub it with the dry salt; this will cause a slimy lather to appear that should be rinsed off whenever it becomes bothersome. Continue to rub it with salt until no more lather appears and the skin begins tearing off in strips. Remove the skin, cut the tentacles into convenient-sized strips, and your Octopus is ready to cook. A good way is to boil the tentacles in very little water for about 30 minutes, then cut them into bite-sized pieces. Melt a cup of butter, add 2 crushed cloves of garlic and one tablespoon finely chopped parsley. Dip each bite in the sauce as it is eaten.

An Oriental method is to slice the flesh into thin pieces and fry it in hot peanut oil until it begins to curl. Eat with a sauce made of Chinese mustard, grated green ginger, and soy sauce combined to suit your own taste.

The Hawaiians eat Octopus raw, and, far from being horrible, it is at least as good as the raw Cherrystone Clams we relish. Cut the meat into small bite-size pieces and mix with it 1 chopped onion and a few finely crushed chili peppers. Put it in the refrigerator for about an hour, so the flavors will permeate the meat, then serve it on toothpicks as hors d'oeuvres or cocktail snacks. The Hawaiians eat it with *poi*, but that is a delicacy unobtainable here.

When the cleaned meat is run through a food chopper, the resulting Octopus-burger can be used in any of the ways recommended for preparing ground-up whelk, clam, or other tough shellfish. It is exceedingly good heaped on a scallop shell, covered with a slice of bacon, and baked until the bacon is nice and crisp. No matter how it is prepared, I'll wager that from the first bite you will cease to consider this creature horrible and ugly.

I had not intended to mention that the heart of the Palmetto, or Cabbage Palm, of the South is one of the finest vegetables and salads in the plant kingdom, for in order to get this delicacy one must destroy this beautiful palm. However, on a recent trip through the South, I observed many Palmettos being cut down to make room for new superhighways, supermarkets, supermotels, and parking lots. Since these palms were being destroyed anyway, I determined that at least some of the delicious palm hearts would not go to waste, and gathered as many as I could use from the felled trees.

The Palmetto is easily recognized, because it is usually the only native palm in the area where it grows. From North Carolina to Florida and around the Gulf Coast, it often covers mile after mile of the sandy soil back of the beaches. Usually shrubby, it sometimes grows 20 or 30 feet high and has been known to reach a height of more than 50 feet, with a trunk diameter of over 1 foot. The delicious heart, vulgarly and insultingly called a "cabbage," is really the terminal bud found in the center of the leaf cluster. It is composed of tightly folded, unborn leaves, and is snow-white, crisp, tender, sweet, and delicious.

Palm heart can be cooked and eaten like cabbage, but you will find it far better than that plebeian vegetable. Thinly sliced, chopped, or shredded, it makes a wonderful addition to almost any tossed salad. Even better, serve shredded palm heart alone with your favorite dressing. One cup of shredded palm heart mixed with 1 cup of claw meat from Stone Crabs, topped with a clear French dressing, and mounded into halves of really ripe and soft avocados, makes a dish that is the very highest expression of the saladmaker's art.

Another plant of the southern seacoasts that came to my attention was the shrubby Yaupon, or Cassina, a seaside relative of our common Christmas holly. Its leaves make the finest wild-plant tea available in North America. This tea actually contains appreciable amounts of caffeine, making it slightly stimulating like Oriental tea or coffee. Some Southern Indian tribes knew the secret of making this bracing beverage, but these same Indians, at some of their solemn ceremonies, used another beverage, the famous "black drink," which acted as a powerful emetic. The black drink was a concoction of various roots and herbs, but some careless observer once reported that it was the Yaupon tea that had this undesirable effect, thus giving this valuable plant an undeserved bad reputation. This error has persisted in the literature, and one can still read current accounts that connect the Yaupon with the infamous "black drink." Because of this error, it was given the scientific name it bears, *Ilex vomitoria*. It is time we correct this ancient error and clear the name of this fine seaside plant, for it makes a mildly stimulating tea that is delicious and as wholesome as the commercial tea and coffee we buy. It is a close relative of *maté* or Paraguay tea, which is highly appreciated in South America, has been domesticated, and is commonly sold commercially.

Our own Yaupon is a small shrub, 1 to 3 feet high, stiffly branched, and often forming small thickets back of the beach from Virginia to Florida and around the Gulf Coast. It is an evergreen, and can be gathered any time of the year. The leaves are oval and small, about 1 inch long, and they are bluntly scalloped on the edges. In the autumn it bears small red berries resembling those of the related hollies. I tried making tea of the green leaves, but found it herby and unpleasant. Just dried in the shade it was little better, but when I put a large bake pan full of the leaves in an oven, turned the heat very low, propped open the oven door, and roasted them until they were dry and crumbly, they made very fine-tasting tea with a definitely stimulating effect. Crumble the roasted leaves, remove the stems and mid-veins, and make the tea exactly as you make ordinary Oriental tea. In Paraguay, they make it in a bowl and drink it through a glass tube that has a strainer on the lower end of it, to keep from getting the floating leaves in their mouths, but I found a fine kitchen strainer worked just as well and allows one to drink the tea from ordinary teacups. Even better, put the crushed leaves in a perforated metal tea ball and immerse this in the boiling water. Use about as much of the leaves as you would of regular tea to get a beverage of comparable strength. It is the perfect drink to accompany a seafood meal for which you have foraged the food.

One could go on and on. There are the sea cucumbers, not plants, but relatives of the starfishes and Sea Urchins. These aberrant creatures are easy prey for the tide-pool hunter and diver, and when dried they are the famous *bêche de mer*, or *trepang*, from which Chinese chefs make epicurean soups and stew. There are the conchs and helmet shells of our Southland, both excellent food mollusks and large enough to be worthwhile. There are hundreds of kinds of inshore fishes, easily caught from bridges, piers, or the banks of tidal streams, that have not been mentioned. There are many other seaside plants that furnish possible food for man. There are dozens of recipes that it would be fun to discuss, such as conch chowder, grits and grunts, trepang soup, and fried helmet steaks.

It is my hope that this book will open new dimensions of seaside vacationing to those who formerly have thought of the ocean chiefly as a place to swim. Ocean bathing on a smooth beach is a rare

Yaupon
(*Ilex vomitoria*).

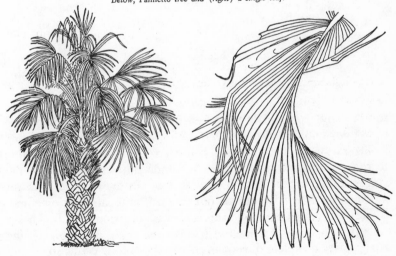

Below, Palmetto *tree and* (right) *a single leaf.*

pleasure, but it is not the only joy the ocean can give us. There are thousands of miles of tidal shoreline that are unsuited to ocean bathing but ideal for this kind of food-gathering fun. Among the best are the deeply indented and island-studded shores of Maine, New Brunswick, Nova Scotia, Newfoundland, eastern Quebec, and Labrador. On the West Coast there are thousands of uninhabited miles of spectacularly beautiful shores in British Columbia and southeastern Alaska, where mountains, forest, and sea compete for your attention. In Lower California, one can find desert scenery and marine landscapes side by side, with shorelines loaded with many delicious edible creatures. There is the marvelous coast of the Gulf of Mexico and the weird delta islands at the mouth of the mighty Mississippi. The Sea Isles of Georgia and South Carolina offer splendid opportunities for unique seaside camping vacations. Even the commercialized and developed Central Atlantic States, with their great resorts, offer much more than the average vacationer sees. By turning your back on the sandy, outer beaches, and wandering inland toward the maze of saltwater bays, inlets, estuaries, tidal streams, and salt marshes, with their flourishing life-forms, you can glean many an epicurean meal. All these states and provinces will welcome you if you will take the trouble to find out about their laws and regulations before you start helping yourself to their seaside bounty. Usually the laws are generous, if you are gathering the seafood strictly for your own use.

If you are dismayed at the task of learning to recognize so many strange creatures, you needn't be. Getting to know these life-forms is a joy in itself, and you don't need to do it all at once. It is not necessary that you be able to identify every creature of the sea and shore before you begin this food-gathering fun. As soon as you can certainly recognize only one species, you are ready to start gathering seafood, and you will be surprised how your acquaintance among the shore animals will grow. One of the pleasantest ways of learning to identify the edible mollusks is to start a shell collection. Adding to this collection can contribute still more interest and joy to the family seaside vacation. Soon every member of the family will be on a first-name basis with dozens of beautiful, unusual, and sometimes edible creatures that none of you ever heard of before. Dr. R. Tucker Abbott, in his pocket edition of *How to Know the American Marine Shells*, gives complete directions on how to collect, clean, store, and identify all the mollusks mentioned in this book and many more.

I hope I have made it clear that the best way to enjoy this pastime is not to follow directions slavishly, either mine or those of others. To wring the maximum amount of pleasure from this hobby you must improvise, explore, experiment, and invent. Maybe we will meet some day along a rocky shore or in a salt marsh, and there exchange techniques, methods, and recipes. I hope so.

Index

Index